OPTOMECHANICAL SYSTEMS ENGINEERING

WILEY SERIES IN PURE AND APPLIED OPTICS

Founded by Stanley S. Ballard, University of Florida

EDITOR: Glenn Boreman, University of North Carolina at Charlotte

A complete list of the titles in this series appears at the end of this volume.

OPTOMECHANICAL SYSTEMS ENGINEERING

KEITH J. KASUNIC
Optical Systems Group, LLC
University of California at Irvine
Georgia Tech - Distance Learning and Professional Education

WILEY

Published by John Wiley & Sons, Inc., Hoboken, New Jersey
Published simultaneously in Canada

For general information on our other products and services or for technical support, please contact our Customer Care Department within the United States at (800) 762-2974, outside the United States at (317) 572-3993 or fax (317) 572-4002.

Wiley also publishes its books in a variety of electronic formats. Some content that appears in print may not be available in electronic formats. For more information about Wiley products, visit our web site at www.wiley.com.

Library of Congress Cataloging-in-Publication Data:

Kasunic, Keith J.
  Optomechanical systems engineering / Keith J. Kasunic.
    pages cm
  Includes bibliographical references and index.
  ISBN 978-1-118-80932-7 (cloth)
1. Optical engineering.   2. Optoelectronic devices.   I. Title.
  TA1520.K38 2015
  621.36'9–dc23
                                        2014007285

Set in 10.5/13pt Times by SPi Publisher Services, Pondicherry, India

10 9 8 7 6 5 4 3 2 1

1   2015

CONTENTS

PREFACE

A great mechanical engineer's experience producing camshafts rarely translates into excellence in opto-mechanical engineering.

John Lester Miller
Principles of Infrared Technology

Most of the literature on optomechanical design takes an in-depth look at the many important details needed to develop optical hardware that works—details such as the specifics of alignment mechanisms, screw threads, epoxies, and the like. While useful, this approach also makes it difficult for readers to understand the basic principles behind the designs. Why, for example, is it more difficult to maintain alignment with a fast, off-axis mirror? Or when does the heat from a high-power laser need to be removed with liquid coolant rather than natural air convection? Or is it necessary that a 3-point kinematic mount be used in all situations, or are there designs where a semi-kinematic mount is perfectly acceptable?

This book takes a step back from the detail-level considerations and instead covers the fundamental principles on which optomechanical design rests. It is unique in that the approach taken in the current books—for example, encyclopedic reviews of the many optical instruments that have been built over the years or collections of equations which only those with sufficient background in mechanical engineering can understand—is not used. Instead, it first reviews the basic concepts of optical engineering, showing how the requirements on the optical system flow down to those on the optomechanical design.

The concepts and equations needed to design the optomechanical system then follow naturally. While it does utilize case studies as a pedagogical tool, the cases illustrate fundamental principles, not encyclopedic reviews. This, then, is a book on optomechanical *systems* engineering—a prerequisite for the many details of optomechanical design found in other books.

While this book may seem to be intended only for mechanical engineers, it is also written for the broader audience of engineers, scientists, managers, and technicians with little or no background in mechanical engineering. This includes anyone designing optical, electro-optical, infrared, or laser hardware. In addition to mechanical engineers, a typical reader might be an optical engineer who wants to learn the fundamental principles of optomechanical engineering, as well as electrical, systems, and project engineers who need to do the same but do not have the academic background consisting of courses in mechanics, strength of materials, structural dynamics, thermal management, heat transfer, and kinematics. Even mechanical engineers with training and experience in these disciplines will find this book useful, as Mr. Miller's comment at the beginning of this preface indicates.

The book begins with an overview of optical engineering, illustrating how the requirements on the optical system determine what the optomechanical system must be able to do. The overview includes optical fundamentals (Chapter 2) as well as the fabrication (Chapter 3) and alignment (Chapter 4) of optical components such as lenses and mirrors. From there, the concepts of optomechanical engineering are applied to the design of optical systems, including the structural design of mechanical and optical components (Chapters 5 and 6), structural dynamics (Chapter 7), thermal design (Chapter 8), and kinematic design (Chapter 9). Finally, the book closes with a summary chapter tying everything together from a systems engineer's perspective (Chapter 10).

Tucson, Arizona   KEITH J. KASUNIC

1

INTRODUCTION

Despite our best efforts and intentions, Mr. Murphy—a rather jovial fellow who happens to be a bit sensitive to these things, but who also has a fondness for *Schadenfreude*—will remind anyone developing optical hardware of his inescapable Law, whether they like it or not. For those who are not prepared, the reminder will be unexpected, and schedules, budgets, and careers will eventually be broken; for those who are prepared, his reminder will be much less painful—and soon forgotten, as the customer's happiness at receiving working hardware reminds us why we are engineers in the first place.

Fortunately, Pasteur's antidote to Murphy's near-deadly "snakebites"— that chance favors the prepared mind—is our opportunity to remove most of the fatalism from engineering. The type of preparation that this book provides goes under the name of *optomechanical engineering*—an area of optical systems engineering where "the rubber meets the road," and thus has the highest visibility to managers and customers alike [1].

It is sometimes said that optomechanical design is a relatively new field, but the truth is a bit more complicated. Not surprisingly, it is as old as optical engineering itself, a field that dates back to at least the early 1600s when the Dutch were assembling telescopes.[1] Notable contributions by people who

---

[1] The early 1600 date given in this chapter is that when theory was first turned to engineering practice in the form of complex optical instruments such as telescopes.

Optomechanical Systems Engineering, First Edition. Keith J. Kasunic.
© 2015 John Wiley & Sons, Inc. Published 2015 by John Wiley & Sons, Inc.

are otherwise known as great scientists—but should also be recognized as optomechanical engineers—include:

- Galileo—While he did not invent the telescope, a practical telescope architecture using refractive (lens) components is named after him.
- Isaac Newton—To develop his theories on the nature of light, he invented the first practical telescope using reflective (mirror) components.
- James Clerk Maxwell—In addition to his brilliant discovery of the electromagnetic nature of light, he developed a structural theory of trusses and experimented with photoelasticity and kinematic mounts.
- Joseph Fourier—He made many contributions in the areas of heat transfer and thermal design, including the discovery of Fourier's law of conduction and the Fourier transform for analyzing vibrations, heat-transfer problems, and more recently, electrical circuits.
- William Thomson (aka Lord Kelvin)—He is best known for his work in thermodynamics, developing the Kelvin temperature scale. In addition, he deserves to be recognized for his work in kinematics, inventing the Kelvin kinematic mount.

In short, these were people who were trying to solve difficult physics problems but were unable to do so until they first solved the instrumentation problems of how to make an apparatus stiff, stable, repeatable, and so on, that is, solve the state-of-the-art optomechanical problems.

More recently, we are still trying to solve difficult problems including the following applications and even quantum optomechanics [2]:

- Aerospace: infrared cameras, spectrometers, high-power laser systems, etc.
- Biomedical: fluorescence microscopy, flow cytometry, DNA sequencing, etc.
- Manufacturing: machine vision, laser cutting and drilling, etc.

In the following sections of this chapter, we first take a look at what a typical optomechanical system might consist of (Section 1.1), the skills needed to engineer such a system (Section 1.2), and the mindset needed to do this efficiently (Section 1.3).

1.1 OPTOMECHANICAL SYSTEMS

If we buy an optomechanical system today, what would we expect to get for our money? Figure 1.1 illustrates a complex biomedical product known as a swept-field confocal microscope—a microscope with some unique features that allow it to image over a wide field-of-view with high resolution [3, 4].

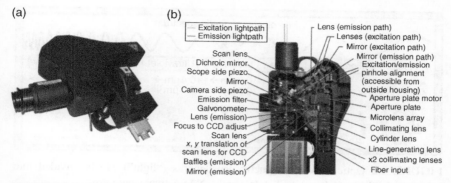

FIGURE 1.1 A complex optical system such as a swept-field confocal microscope requires a large number of optomechanical components packaged into a small volume. Credit: LOCI and Laser Focus World, Vol. 46, No. 3 (Mar. 2010) [3].

Given its complexity, the designers of this microscope had to struggle with many issues that are not obvious to the eye, including:

- Assembly and alignment—Can the optical components all be assembled in a small package and maintain critical alignments such as the "Focus to CCD" distance (for which an adjustment is provided)?

- Structural design—Are the overall structure and the optical submounts stiff enough to keep things in alignment due to self-weight or shock loading?

- Vibration design—Have scan mirror vibrations been isolated from the optics and prevented from causing the optics to move ever so slightly (but more than is acceptable)?

- Thermal design—With components such as the piezos and galvanometers dissipating heat in such a small volume, is there even enough surface area to transfer this heat without the external box temperature getting excessively hot?

- Kinematic design—If the microscope needs to be repaired, is there a way to remove critical optics that allows them to be replaced in the field, without a major realignment at the factory?

- System design—Have all the interactions between the elements been considered, for example, the effects of heat on the optics?

Before getting to these topics, it is important to first understand that common to all optomechanical systems is the use of electromagnetic waves known as "light" (Fig. 1.2). This refers to the wavelength of these waves—on the order of 1 μm, but extending down to 0.1 μm or so and up to ~30 μm—and distinct from "radio" waves, with much longer wavelengths. Controlling the curvature and direction of optical wavefronts with lenses and mirrors is what

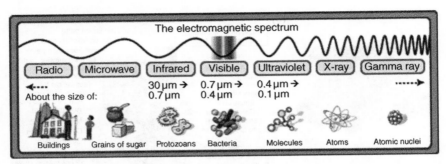

FIGURE 1.2 Optical electromagnetic wavelengths ("light") can be divided into infrared, visible, and ultraviolet bands. Credit: NASA (www.nasa.gov).

allows us to create optical images, or determine the wavelengths present, or measure the power transported by a wavefront. Keeping mechanical parts aligned and stable to <1 µm—an extremely small dimension equal to ~40 micro-inches (or 0.04 milli-inches, often pronounced in abbreviated form as "mils")—is one of the many challenges of optomechanical engineering.

1.2 OPTOMECHANICAL ENGINEERING

So an optomechanical system has a few lenses and a detector—how difficult can this be to design? As we have just mentioned, the small size of an optical wavelength—and thus the mechanical accuracy required—is one of the difficulties. Paul Yoder and Dan Vukobratovich have published the majority of recent books showing us many of the implementation difficulties, and have many useful details and hints on building hardware [5–10]. Even "just" a packaging job is not straightforward, for example, as many laser jocks have discovered when trying to convert their laboratory hardware into a commercial product (Fig. 1.3).

An example of one of the steps required for the transition from lab to marketplace—or even optical designer's desk to working prototype—is shown in Figure 1.4. Here, a lens designer has determined that a three-lens system is required to meet the customer's needs (or "requirements"). The lens design-er's deliverable to the optomechanical engineer is an optical prescription from lens design software such as Zemax or Code V, consisting of lens geometries (surface radii and diameter), materials, and spacings between the lenses.

The lens designer must also provide a tolerance analysis showing how sensitive each lens is to misalignments such as tilt, centration, and spacing errors—and how much of each is allowed. In addition, the lens designer must deliver a fabrication analysis for each lens, specifying the tolerances on the lens surfaces, refractive indices of the lenses, and so on—a topic we will look

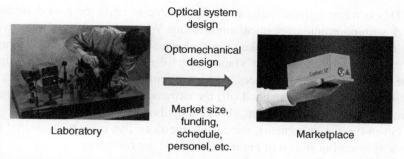

Optical system
design

Optomechanical
design

Market size,
funding,
schedule,
personel, etc.

Laboratory

Marketplace

FIGURE 1.3 The transition from laboratory to marketplace is critically dependent on the skills of the optomechanical engineers. Photo credits: Permission to use granted by Newport Corporation; all rights reserved.

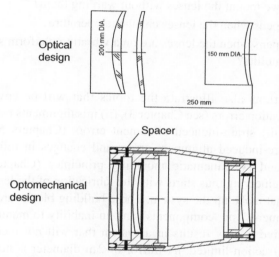

Optical
design

200 mm DIA.

150 mm DIA.

250 mm

Spacer

Optomechanical
design

FIGURE 1.4 An example illustrating the steps required to move from an optical design prescription to a complete optomechanical design. Credit: G. E. Jones, "High Performance Lens Mounting," Proc. SPIE, Vol. 73, pp. 9–17 (1975).

at in detail in Chapter 3. Ideally, the optical prescription, alignment analysis, and fabrication analysis must be developed in coordination with the optomechanical engineer; in practice, this is not always done, with the consequences illustrated with an example in Section 1.3.

Given these inputs from the lens designer, what the optomechanical engineer must then determine is as follows: are the fabrication and alignment tolerances (i.e., allowable variations) feasible, given the manufacturing, technical, environmental, cost, and schedule constraints? That is, is it possible to take the lens designer's prescription and convert it into manufacturable hardware such as that shown in the bottom graphic of Figure 1.4?

This is never a trivial task, and it is possible to undo the lens designer's work with poor optomechanical engineering. What makes the task nontrivial are the requirements that make optomechanical engineering so challenging: (i) the extremely small ("tight") fabrication tolerances, (ii) the extremely tight assembly (alignment) tolerances, perhaps <1 μm for displacements and 1 μrad for angular misalignments, and (iii) the extreme sensitivity to environmental conditions such as temperature, shock, and vibration.

Given these requirements, the design questions that must be answered for the lens assembly shown in Figure 1.4 include the following:

- Can we make (or buy) the optics to the quality specified by the lens designer?
- How can we mount the lenses without warping them?
- What happens when the lenses change temperature?
- What happens when the lenses are on a vibrating platform such as a car, boat, or satellite?

These questions also illustrate the topics that will be covered in this book: (i) fabrication errors (see Chapter 3), (ii) misalignments between lenses (Chapter 4), (iii) strain-induced alignment errors (Chapters 5, 6, and 7), (iv) temperature-induced alignment errors and changes in refractive index (Chapter 8), and (v) kinematic mounting principles (Chapter 9). Optomechanical engineering thus starts with the fabrication of the lenses, mirrors, and so forth that make up the system. These building blocks are collectively known as "elements" or "components" and an inability to manufacture them with the required quality results in a system that will not meet specs. For example, a diffraction-limited, $f/1$ mirror of 2 m diameter is not straightforward to build, and fabrication technology may be the limiting factor in its optical performance. Yet if the mirror needs only to collect photons and need not generate good image quality, then a large-diameter optic becomes more feasible. The ability to meet specs thus depends in part on how well the image quality requirements of the system match up with the optical quality available from the fabrication process.

After fabrication, optical elements are assembled into an instrument or system; this requires that they be accurately aligned with respect to each other to insure crisp, high-resolution image quality. The distance between the primary and secondary mirrors in the Hubble Space Telescope, for example, was ~5 m—to a tolerance of ±1.5 μ, or 1 part in 1.7 million! Typical alignment problems for an optical system include: (i) the tilt angle with respect to either of the two axes perpendicular to the optical axis; (ii) decenter, or displacements along either of those two axes; (iii) despace, or the incorrect placement of one element with respect to another as measured along the optical axis; and (iv) defocus, a specific type of despace pertaining to the placement of the detector with respect to the imaging optics as measured along the optical axis. These assembly alignments and their effects on optical performance are reviewed in Chapter 4.

The most critical factors determining whether or not an optical system can be built are the fabrication and assembly tolerances; these are the acceptable variations of the fabrication and assembly parameters. Tolerances are determined by how sensitive performance is to small changes in system-level parameters such as f/#. This sensitivity flows down to component-level tolerances through the dependence of f/# (for instance) on aperture and effective focal length (EFL); EFL, in turn, depends on the refractive indices, surface curvatures, and thickness and spacings of the lenses in the system. An EFL that is extremely sensitive to changes in index, for example, is considered difficult to fabricate; similarly, a lens-to-lens spacing that is highly sensitive to misalignments is difficult to assemble.

Moreover, tight assembly tolerances imply that the optics are also extremely sensitive to vibration and changes in temperature. Once aligned, an optical system will usually remain so on a vibration-isolation table in a climate-controlled building; however, the system may become useless once it is moved out into the field, where environmental effects—including temperature changes, shock, vibration, humidity, ocean spray, vacuum, and radiation—become important. The two most common problems are temperature and vibration, either of which can quickly destroy the instrument's ability to meet a high-resolution image-quality spec. Even systems that are precisely fabricated and accurately aligned can fail to meet image-quality requirements when, for example, the temperature drops below freezing or the instrument has been bouncing around in the back of a truck during a bumpy off-road trek through the Colorado mountains.

In short, *better design performance does not necessarily translate into a better optical system*. Many other factors affect overall performance, including fabrication, alignment, and environmental influences such as temperature and vibration. The sensitivity to fabrication tolerances and misalignments is the key determinant of cost and producibility. The best design is therefore one that has large tolerances in all areas.

Optomechanical systems consist, of course, of more than lens assemblies; an example of a more complex system is shown in Figure 1.5. Here, the authors illustrate the use of hardware "building blocks" previously made for other projects to save development time and money on future projects—a much-needed approach for the aerospace business which the authors are in.

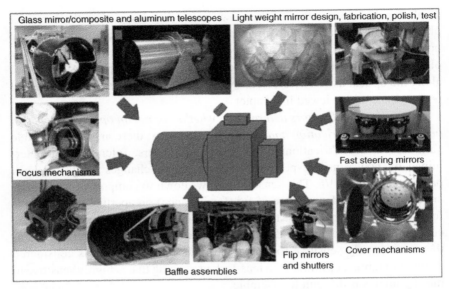

FIGURE 1.5 A complex optomechanical system requires the coordination of many different disciplines and technologies. Credit: Tony Hull and Mark Schwalm, "Spaceborne telescopes on a budget: paradigms for producing high-reliability telescopes, scanners, and EO assemblies using heritage building blocks," Proc. SPIE, Vol. 8044 (2011).

Shown in the figure are a number of optomechanical assemblies and subsystems such as a fast-steering mirror, focus mechanism, and a telescope sub-assembly (lightweight mirrors, metering structure, etc) which are all combined to produce a complex telescope assembly. The optical prescription in this case—that of the primary and secondary mirror—is simpler than that of the lens assembly shown in Figure 1.4; the optomechanical design, on the other hand, is clearly more intertwined with the optical design—the available scan angle for the fast-steering mirror must be consistent with the field-of-view of the optics, for example—illustrating the need for a systems-level approach to optomechanical engineering.

The systems approach to optomechanical engineering is also necessary for the lens assembly shown in Figure 1.4. For example, estimates of athermalization for such a lens often include only the mechanical effects of thermal expansion; also necessary, however, are the optical effects of change in refractive index with temperature. Only the combination of these two—an integrated optomechanical systems design—will produce a system that is relatively insensitive to temperature changes. Thus the appropriate mindset for developing optical hardware must *not* be one of the lens designer saying to the optomechanical engineer "Here's the lens design—go for it!"; instead, a collaborative approach involving all relevant engineering disciplines answering the question "What should the system look like, given the technical, manufacturing, cost, and schedule constraints?" must be used. Section 1.3 illustrates the concept in more detail.

1.3 OPTOMECHANICAL SYSTEMS ENGINEERING

It is often the case that an optomechanical designer will want to jump right into the "epoxies-and-fasteners" (aka "glues-and-screws") details of optomechanical engineering, without first doing technical and cost trades of design options. These trades are required for an overall system approach to the problem, so that even without having the time or budget to do formal trades, it is invaluable to at least take a higher-level look at an initial design from the perspectives of manufacturability, assembly, optical alignment tolerances, and so on.

It takes a certain mindset to do this right, and it is difficult to find people with sufficient technical expertise to know the trades, but not so "detail oriented" at the beginning as to lose sight of the big picture. It is important to get the details right, but it is even more important to get the architecture going in an intelligent direction. As with aiming errors on a long pool table, engineering errors at the beginning of a project propagate to create huge delays in schedule and cost overruns in budget (Fig. 1.6) [11], resulting in missed "shots" that will not impress the customer. In these cases, even heroic amounts of "glues and screws" effort cannot pull a successful story out of the fog of failure.

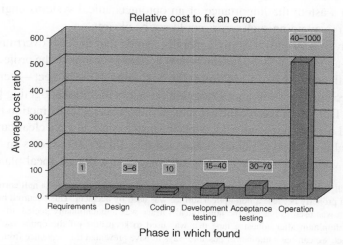

FIGURE 1.6 Engineering errors at the beginning of a program propagate to create huge budget (and schedule) overruns if not corrected early on. Source: NASA Johnson Space Center [11].

Recent highly-visible examples of "missed" optical engineering projects include the Hubble Space Telescope [12], a classified program reported on by *The New York Times* [13], and the James Webb Space Telescope (JWST) [14]. In the case of the Hubble, the lack of end-to-end testing of the telescope's images was the root cause of a $250 million error which required an on-orbit repair mission in space. The technical "glitch" that the end-to-end testing would have uncovered was an error in fabrication of the shape of the primary mirror—an error which itself was caused by a mistake in the assembly of the calibration optics used to measure the shape. This mistake was not so much one of architectural design as it was a Murphy's Law cascade of smaller errors which may or may not have been identified with better optomechanical engineering training and procedures.[2]

The classified program was covered in depth in *The New York Times* [13], where it was disclosed that a major spy satellite program was unable to deliver the performance promised by the aerospace contractor. After $5 billion was spent, the optical system was still not ready for launch. After more delays, what was eventually launched was an investigation into the root cause of the problem; an aerospace executive concluded: "There were a lot of bright young people involved in developing the concept, but they hadn't been involved in manufacturing sophisticated optical systems. It soon became clear the system could not be built." This seems to be a clear case of an architectural error, a fundamental mistake in a basic concept which was only later found to be flawed. Further underlining the architectural error was another statement from an investigator that "The train wreck was predetermined on Day 1." This is, of course, *very* strong language, but strong language is required after $5 billion has been wasted; the importance of an optomechanical systems engineering approach to avoiding such errors could not be clearer.

Finally, it was until recently generally agreed that the cost overruns on the JWST—approximately $9 billion is now needed to complete the project, compared with an estimated $2 billion at the beginning of the project—were based not on technical performance issues but on errors in initial budgeting [14]. As the JWST program has recently discovered, however, the distinction between technical performance and budget performance is not always clear cut, as the two are almost always entangled. For example, Figure 1.7a shows an imaging camera, with three lenses creating an image on a digital focal-plane array

---

[2] Specifically, the instrument used to measure the shape of the primary mirror (a null corrector) was not aligned properly because a non-optical surface was used by accident. This occurred because the paint which was used to blacken the non-optical surface was unintentionally peeled off—exposing reflective aluminum after adhesive tape that was used to keep dust off the optics was removed. In retrospect, we can see a number of possible ways to have prevented this sequence from becoming catastrophic, such as making end-to-end measurements of the system performance to check for such errors. See Ref. [12] for more details.

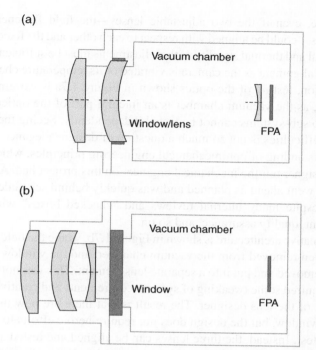

FIGURE 1.7 In (a), a lens design with an external lens, curved window, and field flattener is extremely difficult to align. In (b), all lenses are combined into a compact mount to ease alignment, assembly, and test.

(FPA). The FPA is in a vacuum chamber, as is the lens located near it (known as a field flattener; see Ref. [1] for more details). In addition, the lens designer—in an effort to reduce the parts count—has combined the function of a lens and the vacuum window by putting curvature in the window itself, thus reducing the number of parts by one lens. On this basis, a cost estimate was made for a potential customer—including the cost "savings" from combining optical elements—who agreed to fund the project.

While it seems reasonable to combine elements like this to reduce the parts count, what is found instead is that this design is not manufacturable, and the cost estimate was quickly exceeded. Specifically, the accumulation (or "stackup") of tolerances on the many individual mechanical parts required an extensive assembly and alignment procedure. As we will see in Chapter 4, the lenses must all be aligned to a certain degree with respect to each other in terms of their spacings, tilt angles, and centering. This is next to impossible for the design shown in Figure 1.7a, given the large separations of the optics, the individual parts on which they are mounted, the mechanical paths connecting the parts, and the tight spacing and centration tolerances this design required.

Furthermore, even if the two adjustable lenses—the field flattener and the external lens—could be aligned with respect to each other and the fixed window, the structural and thermal stability of the adjustments must be sufficiently robust to maintain alignment as the camera is vibrated or its temperature changes.

In addition, testing of the optics shown in Figure 1.7a is extremely time-consuming, as the vacuum chamber is an integral part of the optical design, and the two sub-systems cannot be tested independently. Seeing these manufacturing difficulties is not so much a question of design "elegance" as it is a basic understanding of optomechanical engineering principles, which neither the lens designer nor the mechanical engineer on this project had. As a result, the project went ahead as planned and was quickly behind schedule and over budget—despite many internal reviews and "checked boxes" with project management, quality assurance, and so on.

An alternative architecture is shown in Figure 1.7b, where the field flattening lens has been removed from the vacuum chamber, and the window curvature has been removed and put into a separate lens element. This was not a difficult task but required some tweaking of system requirements and creative thinking on the part of the lens designer. The result is that there are now three lenses and a flat window, but the design does not require heroic efforts to assemble, align, and test. Instead, the three lenses can be aligned and tested as a unit in a compact mount and then bolted to the vacuum chamber for final positioning of the image on the FPA. In addition, the structural and thermal stability is much easier to maintain over the smaller package size, as we will see in Chapters 5, 6, 7, 8, and 9.

So based on the optomechanical system architecture shown in Figure 1.7a—which did not take basic manufacturing and alignment trades into account—the budget was proven to be grossly underestimated; the much lower-cost estimate for the design shown in Figure 1.7b, on the other hand, was not exceeded, and this version of the camera is in use today. So is the cost over-run for the Figure 1.7a design a problem of errors in initial budgeting—or instead one of poor engineering? In Figure 1.7a, the mechanical engineer did not have the required optomechanical skills to assess manufacturability; in other cases, the optomechanical engineers are not included in the discussion at the beginning and are left to salvage whatever they can from the optical design that is given to them after the project is funded. In both situations, the budget problem can be traced back to the lack of utilization of the technical skills required to get the job done.

The most important technical skill is that of system design, and from this perspective, "collaborative" is more than a buzzword to be used on employee annual review forms. Instead, it is how a *system* is designed—as distinct from a component, where an individual engineer has pretty much all the information they need to sit down and design parts to the point where they can be fabricated.

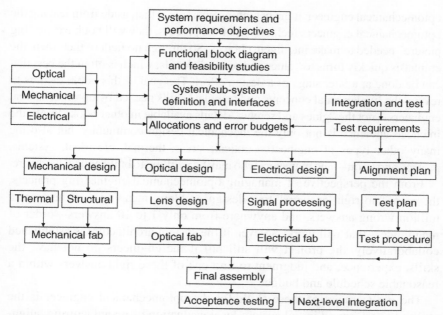

FIGURE 1.8 Optomechanical engineers must be involved at both the system architecture (top left) and detail design (middle left) phases of product development. Adapted from Keith J. Kasunic, *Optical Systems Engineering*, McGraw-Hill (2011).

A system, on the other hand, requires the knowledge and experience of many people with widely varying skills and backgrounds—optical engineers, mechanical engineers, electrical engineers, software engineers, manufacturing engineers, project engineers, systems engineers, and project managers (Fig. 1.8). It does not require large meetings with all these people on a daily basis, but it does require coordination—that is, collaboration—between the key interfaces such as the optical and optomechanical engineers (as we will see in detail in this book), or optical and electrical engineers (for an imaging system which uses digital FPAs, for example), or the many other interfaces that the systems engineer is responsible for [1].

Summarizing, a problem that has caused cost and schedule overruns on many projects is that the entire system design is disjoint from the beginning. That is, the optical engineers work to design a lens based on requirements given to them by the systems engineer, and then throw the lens design "over the wall," that is, without inputs from the optomechanical engineer as to fabrication requirements, alignment tolerances, vibration instabilities, thermal drift, and so on.

System architecture is the most important thing to get right, and the optomechanical system is as important as the others [1]. Not including the optomechanical engineer at the beginning guarantees that, aside from leaving the optomechanical engineers feel left out of the process, they will not have the "big picture" needed to make intelligent choices. So through no fault of their own, the emphasis quickly turns to "glues and screws" details, which is often the best that can be done at a later stage of a design effort. The goal of this book, then, is to review the fundamental concepts and technical skills needed by optomechanical engineers—not the "glues and screws" details available in other books [5–9]; the intended audience is not only the mechanical design community, but also the many other types of engineers—optical, stress, thermal, electrical, systems, project, and so on—and managers who may be working to build optical hardware.

From the perspective of managing optomechanical engineering projects, there are "no right answers" to a design problem, meaning that there are a million wrong answers, and anywhere from only 1 to 10 answers—order of magnitude—that will work. So even if the system architecture is established collaboratively, the project will still fail if the engineers do not have the skills, experience, and judgment to find one of these right answers within a reasonable schedule and budget.

The most crucial of these skills for the optomechanical engineer is the ability to perform an initial system-level evaluation of manufacturing, alignment, and assembly before investing in detailed performance models or calculations. It is impossible to include everything in this assessment, so the purpose of the analysis and calculations is to do as much up-front work as is practical within the time and budget constraints. As with all trades, it is difficult to maintain a balance between these expenses, and it is usually best to build hardware as soon as possible.

Finally, it is easy to fall into the mistake of thinking that management processes for monitoring progress will insure an on-time, on-budget project that meets all of its technical goals. In the experience of former NASA Administrator Michael Griffin, however, systems engineering processes "… do not help to distinguish a good design from a poor one, nor can they make a poor design better" [15]. Management checklists are important, but technical skills are even more so. As Griffin summarizes in a panel discussion on his experiences: "In engineering, like flying, you need to follow the checklist, but you also need to know how to engineer" [16].

Along these lines, this book starts with an overview of optical engineering, to show how the requirements on the optical system determine what the optomechanical system must do. The overview includes optical fundamentals (Chapter 2) and the fabrication (Chapter 3) and alignment (Chapter 4) of optical components. From there, the fundamental principles of optomechanical engineering are applied to the design of optical systems, including the structural design of mechanical and optical components (Chapters 5 and 6), structural dynamics (Chapter 7), thermal design (Chapter 8), and kinematic design (Chapter 9). The book closes with a summary chapter tying everything together from a system—but not a process—perspective (Chapter 10).

REFERENCES

1. K. J. Kasunic, *Optical Systems Engineering*, New York: McGraw-Hill (2011).

2. M. Aspelmeyer, S. Gröblacher, K. Hammerer, and N. Kiesel, "Quantum optomechanics: throwing a glance," J. Opt. Soc. Am. B, Vol. 27, No. 6, pp. A189–A187 (2010).

3. B. Vogt, L. Yan, M. Szulczewski, and K. Eliceiri, "Swept field confocal overcomes point-scanning microscopy limitations," Laser Focus World, Vol. 46, No. 3, p. 55 (2010).

4. R. Liang (Ed.), *Biomedical Optical Imaging Technologies*, New York: Springer-Verlag (2013).

5. R. E. Fischer, B. Tadic-Galeb, and P. R. Yoder, Jr., *Optical System Design* (2nd Edition), New York: McGraw-Hill (www.mcgraw-hill.com) (2008), Chaps. 16–18.

6. P. R. Yoder, Jr., *Opto-Mechanical Systems Design* (3rd Edition), Boca Raton: CRC Press (www.crcpress.com) (2005).

7. P. R. Yoder, Jr., "Mounting optical components," in M. Bass, E. W. Van Stryland, D. R. Williams, and W. L. Wolfe (Eds.), *Handbook of Optics* (2nd Edition), Vol. 1, New York: McGraw-Hill (www.mcgraw-hill.com) (1995), Chap. 37.

8. P. R. Yoder, Jr., *Mounting Optics in Optical Instruments* (2nd Edition), Bellingham: SPIE Press (www.spie.org) (2008).

9. D. Vukobratovich, *Introduction to Optomechanical Design*, SPIE Short Course Notes (www.spie.org) (2009).

10. D. Vukobratovich, "Optomechanical systems design," in J. S. Acetta and D. L. Shoemaker (Eds.), *The Infrared and Electro-Optical Systems Handbook*, Vol. 4, Bellingham: SPIE Press (www.spie.org), Chap. 3.

11. "Error Cost Escalation through the Project Life Cycle," NASA Johnson Space Center (2004), ntrs.nasa.gov/archive/nasa/casi.ntrs.nasa.gov/20100036670.pdf. Accessed January 13, 2015.

12. J. Hecht, "Saving hubble," *Optics and Photonics News*, March 2013, pp. 42–49.

13. P. Taubman, "Death of a spy satellite program," *The New York Times*, November 11, 2007, p. 1.

14. A. Svitak, "Technical, cost issues persist for Webb telescope," *Aviation Week & Space Technology*, July 22, 2013, http://aviationweek.com/awin/technical-cost-issues-persist-webb-telescope. Accessed January 13, 2015.

15. M. D. Griffin, "How do we fix system engineering?," 61st International Astronautical Congress, Paper IAC-10.D1.5.4; September 27–October 1, 2010; Prague, Czech Republic (2010).

16. N. J. Slegers, R. T. Kadish, G. E. Payton, J. Thomas, M. D. Griffin, and D. Dumbacher, "Learning from failure in systems engineering: a panel discussion." Wiley Online Library (2010). doi:10.1002/sys.20195.

2

OPTICAL FUNDAMENTALS

An optomechanical engineer without a basic background in optical engineering is like an alchemist—without an understanding of the "why" and the "how," the result will be no better than a random mixture, a collection of hardware that may or may not do what is required. Similarly, troubleshooting optical systems that are close to working can quickly become painful if a path or direction is taken based on miscommunication—an error that may have been avoided with a better understanding of the optical engineer's perspective on the problem.

Understanding the optical engineer's craft has three components. The first is that of imaging and its errors. Imaging—illustrated in Figure 2.1 as the collection of light with an optical system such as a camera to produce an image of an object such as a distant mountain scene—relies on the use of a concept known as a *wavefront*; the errors in the wavefront that result in low-quality images are—a majority of the time—a result of poor optomechanical design. This chapter reviews both the imaging properties of lenses (Section 2.1) and their errors (Section 2.2).

The second and third components required for understanding the optical engineer's craft are the fabrication of optical elements such as lenses and mirrors and the general requirements on the alignment of these elements; these will be reviewed in Chapters 3 and 4, respectively.

Optomechanical Systems Engineering, First Edition. Keith J. Kasunic.
© 2015 John Wiley & Sons, Inc. Published 2015 by John Wiley & Sons, Inc.

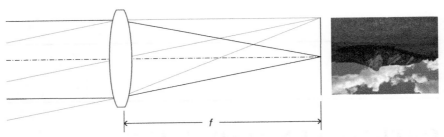

FIGURE 2.1 An inverted image of a distant mountain scene is created using a simple lens with focal length f. Photo credit: Mr. Brian Marotta, Louisville, Colorado.

2.1 GEOMETRICAL OPTICS

By itself, a white card pointed at a scene does not produce an image; instead, lenses and mirrors modify wavefronts to create an image on a card, screen, or detector array. A diverging wavefront, for example, can be converted to a converging wavefront, creating a point-to-point reproduction (i.e., an image) of the object (Fig. 2.2). As we will see in this section, the lens shape and refractive index determine the type and degree of wavefront modification.

The most common function of a lens is to convert planar wavefronts from a distant object into the converging wavefronts that create an image.[1] As illustrated in Figure 2.3, the planar wavefronts are simply a result of the large distance from the object to the lens—given that all wavefronts further from the source have a larger radius R (or less curvature C, since $C = 1/R$). At a sufficient distance from the source, the spherical waves emitted or reflected by the object have a large enough radius to be considered planar (i.e., no curvature, with $C \approx 0$). For optical systems, the wavefronts are spaced by a peak-to-peak wavelength λ on the order of 1 μm, going as low as 0.2 μm (ultraviolet) and as high as 30 μm (infrared).

With planar wavefronts incident on its first surface, the lens is able to convert these wavefronts into an image at a distance past the lens known as the *focal length* (Fig. 2.4a). How the lens is able to do this is governed by Snell's law, whose effectiveness depends on the refractive index of the lens and the radii of its surfaces.

The concept of refractive index (symbol n) indicates how slowly electromagnetic waves move through a lens; a high-index material slows the waves more than a low-index material, in comparison with the fundamental

---

[1] Since the wavefronts from a distant object are almost planar, planar wavefronts are sometimes said to be created by an object "at infinity."

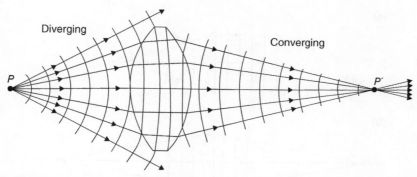

FIGURE 2.2 A lens modifies wavefronts to create an image of the object (point *P*) at point *P'*. Adapted from Warren J. Smith *Modern Optical Engineering*, McGraw-Hill (2008).

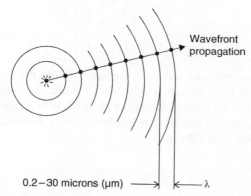

FIGURE 2.3 A pebble dropped in a pond, or an oscillating electron emitting electromagnetic waves, creates wavefronts that propagate with a wavelength λ. Adapted from Warren J. Smith, *Modern Optical Engineering*, McGraw-Hill (2008).

speed of electromagnetic waves in a vacuum (the speed of light c). The velocity of a wavefront propagating through a lens is thus given by $v = c/n$.

Figure 2.5 shows why the surfaces of the lens must be curved to produce an image. By designing an optic which seems to have the general shape of a lens on one side—but without curved surfaces—we see that the wavefronts propagate straight through the "lens." They are thus unable to converge a planar wavefront to an image; instead, the transmitted wavefront has errors which map the surface of the optic. In other words, an optic whose surfaces are both flat—that is, a window—is not able to modify the overall curvature of the incident wavefront.[2]

---

[2] Optics that look flat, but whose shape is modified on the scale of a wavelength—diffractive optics—are exceptions to this rule.

(a)

(b)

(c)

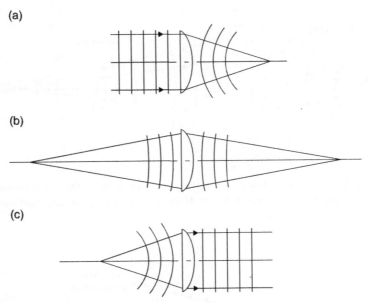

FIGURE 2.4 A lens modifies wavefront curvature. In (a), the wavefront curvature incident on the lens from the left is zero $(R=\infty)$, while the transmitted curvature $C=1/f$. In (b) and (c), the incident wavefront curvature is larger, and the transmitted wavefront curvature is thus smaller. Adapted from Keith J. Kasunic, *Optical Systems Engineering*, McGraw-Hill (2011).

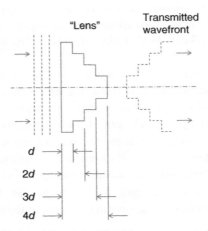

FIGURE 2.5 The surfaces of the optic ("lens") are not curved, and as a result, do not converge wavefronts to produce an image.

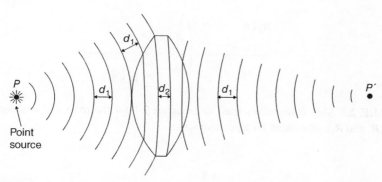

FIGURE 2.6 Changes in wavefront curvature occur when different parts of the wavefront are forced by the curvature of a surface to propagate through materials—for example, air and glass—which have different wavefront velocities. Adapted from Warren J. Smith, *Modern Optical Engineering*, McGraw-Hill (2008).

In contrast, an optic with curved surfaces is able to modify the curvature of the incident wavefront (Fig. 2.6). The reason is that the wavefront exits from the thinner edges of the lens first, where the portion of the wavefront still in the lens, whose index $n \approx 1.5$, is moving more slowly than the portion that has already left the lens and is propagating in air (whose index $n \approx 1$). Because the portion of the wavefront in air is moving faster, it pulls ahead of the portion still in the lens, thus creating curvature of the wavefront if the lens surface itself has curvature.

The shape of the transmitted wavefront depends on the shape of the lens surfaces; ideally, spherical surfaces create spherical wavefronts that can either converge or diverge. The *change* in curvature between the incident and transmitted wavefronts depends on the power of the lens, where power $\Phi = 1/f$ such that a shorter focal length f has more power.[3] For a simple thin lens in air, the power depends on the surface radii (R_1 and R_2), the center thickness CT, and the refractive index n of the material used to make the lens

$$\frac{1}{f} = (n-1)\left[\frac{1}{R_1} - \frac{1}{R_2} + \frac{n-1}{n}\frac{CT}{R_1 R_2}\right] \tag{2.1}$$

Lens power is measured in diopters (units of 1/m), so a lens with a focal length $f = 100\,\text{mm}$ has a power $\Phi = 1/f = 1/0.1\,\text{m} = 10$ diopters. The definition of the radii using the sign convention of Equation 2.1 is given in Figure 2.7. As we will see in Chapter 3, the radii must be fabricated very accurately to give the

---

[3] Lens power (units of diopters) is not the same concept as optical power (units of watts), as collected by an optical system and measured by a detector; see Ref. [1].

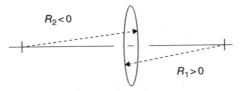

FIGURE 2.7 Sign conventions used in Equation 2.1 for the front and rear-surface radii (R_1 and R_2), illustrated for a positive lens.

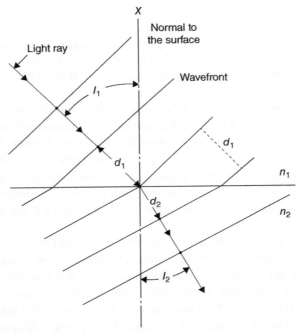

FIGURE 2.8 Snell's law illustrated at a planar surface for $n_2 > n_1$. Adapted from Warren J. Smith, *Modern Optical Engineering*, McGraw-Hill (2008).

correct focal length and very precisely to reduce the surface errors that result in wavefront errors (WFE).

Physically, a larger surface radius (closer to planar) is less effective at bending wavefronts. The reason is Snell's law, illustrated in Figure 2.8 for planar wavefronts incident on a planar surface at an angle I_1 measured with respect to the normal (perpendicular) to the surface.[4] The figure shows that the

---

[4] Snell's law states that the bending angle θ_t depends on the incident angle θ_i and refractive indices (n_i and n_t) on both sides of the surface. Quantitatively, $n_i \sin\theta_i = n_t \sin\theta_t$, which can be approximated as $n_i \theta_i \approx n_t \theta_t$ for $\theta_i \leq 30°$ or so (the near-axis or *paraxial* approximation). This approximation is the reason for using the term *first-order optics*—because of the use of the angle θ to the first power.

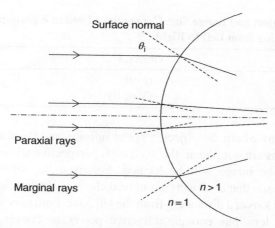

FIGURE 2.9 Because of Snell's law, the lines perpendicular to the surface—or surface normals, shown as dashed lines—determine the angle an incident ray bends.

wavefront is bent (or *refracted*) at the surface into an angle I_2, a result of the difference in refractive indices n_1 and n_2.

A mechanical analogy that is sometimes used to explain this behavior is that of a car traveling on a material with good traction—a paved street, for example—driving at an angle I_1 onto a material with poor traction and large drag (such as loose sand). Because of the difference in car speed between asphalt and sand, the car turns as the right wheel slows down in the sand, while the left wheel does not initially. Similarly, the difference in refractive indices—that is, a difference in the speed of the wavefronts—causes the wavefront to bend with $I_2 < I_1$, given that light moves slower in the second material (the "sand") than it does in the first (i.e., $n_2 > n_1$).

Returning to a lens with curved surfaces, Snell's law applies at each part of the surface, with more bending occurring for the portion of the on-axis wavefront that is farther away from the optical axis. That is, the outer portion of the wavefront—the marginal ray—has a larger incidence angle θ_i with the surface normal than the paraxial ray (Fig. 2.9). The ray that is further from the axis must bend more, if it is to intersect with the paraxial rays to converge at an image location.[5] For the same ray height, steeper surfaces with smaller radii have larger refraction angles; they therefore bend wavefronts more than shallower surfaces with larger radii (less curvature), resulting in a lens with more power.

---

[5] As shown in Section 2.2, the sine-function nonlinearity of Snell's law prevents this from happening perfectly, but for first-order geometrical optics where the marginal rays are relatively close to the optical axis, it is a good approximation.

TABLE 2.1 Object and Image Sign Conventions Used in Equation 2.2 for Light Propagating from Left to Right

Location	Object, s_o	Image, s_i
Left of lens	+ (Real)	− (Virtual)
Right of lens	− (Virtual)	+ (Real)

For situations where the object is not at infinity, and the wavefronts incident on the lens are not planar, the wavefront perspective allows us to understand where the image will be located. For example, referring again to Figure 2.4, we see that as objects are moved closer to the lens, the wavefronts incident on the lens at a distance s_o from the left have more curvature ($C = 1/s_o$). Although the lens has enough refractive power to converge the planar wavefronts to an image distance $s_i = f$ away, it does not have enough power to redirect the diverging spherical wavefronts to the same image distance. As a result, the image distance is farther away from the lens.

Quantitatively, the *change* in curvature between the incident and transmitted wavefronts depends on the power of the lens. This change is summarized by the imaging equation commonly taught in introductory physics courses

$$\frac{1}{f} = \frac{1}{s_o} + \frac{1}{s_i} \tag{2.2}$$

This expression states that, for a lens with a positive focal length f, a smaller distance from an object to the lens ($= s_o$) results in a larger distance from the lens to the image ($= s_i$). Alternatively, the sum of the wavefront curvature incident on the lens ($C_o = 1/s_o$) and transmitted by the lens ($C_i = 1/s_i$) equals the lens power $\Phi = 1/f$.

Because the refractive power of a given lens is a constant, Equation 2.2 shows that the curvature of the object and image wavefronts must sum to a constant (equal to $1/f$). As the object is moved closer to the lens, the wavefronts that fall on it have more curvature; the wavefronts that are transmitted by the lens must then have less curvature and so converge to an image farther away.

The sign conventions given in Table 2.1 lead to a distinction between what are known as *positive* and *negative* lenses. A positive lens has a focal length $f > 0$ and it brings an object at infinity to a converging focus at a positive location (to the right of the lens); a negative lens has a focal length $f < 0$ and it brings an object at infinity to a virtual focus at a negative location (to the left of the lens). Typical shapes for the two basic types of lenses are illustrated in Figure 2.10.

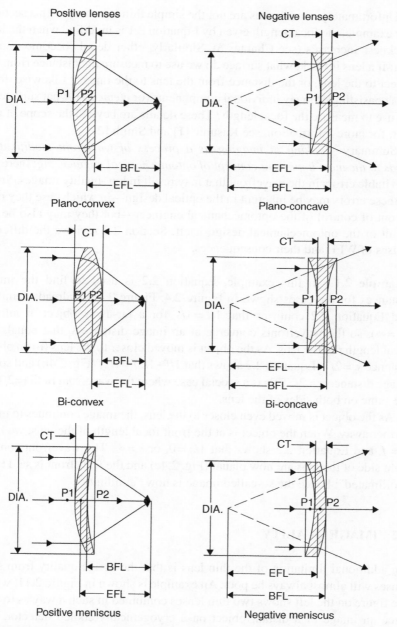

FIGURE 2.10 Three types of positive and negative lenses. Reproduced with permission from Janos Technology LLC.

Unfortunately, most lenses are not the simple thin lens used in this section. For example, the focal length given by Equation 2.1 is modified when the lens thickness increases (see Chapter 3). Similarly, other details become important: If a lens is thick, what surface do we use to measure the distance from the object to the lens, or the distance from the lens to the image? Likewise, for a lens consisting of many individual components (or *elements*), what surface do we use to measure the focal length? These details are beyond the scope of this text; for more information, see Kasunic [1] and Smith [2].

Summary—*looking at imaging as a process of wavefront modification leads to the most important concept of optomechanical engineering*: there are inevitable errors in the wavefront that may result in low-quality images. Some of these errors may be inherent in the optical design—in which case they may be out of control of the optomechanical engineer—but they may also be the result of the optomechanical design itself. Section 2.2 reviews the different causes of WFE and their consequences.

Example 2.1   In this example, Equation 2.2 is used to find the image distances for the cases shown in Figure 2.4. Figure 2.4a is already familiar, and Equation 2.2 confirms that $1/f = 1/s_i$ for a far-away object at infinity $(s_o = \infty)$, so the wavefronts converge at an image distance s_i that equals the focal length f of the lens. As the object is moved closer to the lens (to an object distance $s_o = 2f$), Equation 2.2 shows that $1/f - 1/2f = 1/2f$ (Fig. 2.4b) and so the image distance $s_i = 2f$. This is a special case where the wavefront radii $(=2f)$ are the same on both sides of the lens.

As the object is moved even closer to the lens, the image continues to move further away. When the object is at the front focal length of the lens, we have $s_o = f$, and Equation 2.2 shows that $1/s_i = 0$, or $s_i = \infty$. The wavefronts on the right side of the lens are now planar (Fig. 2.4c) and the wavefront is said to be "collimated"; hence the so-called image is now "at infinity."

2.2 IMAGE QUALITY

An additional limitation of the thin lens is that the image quality from such lenses will almost always be poor. An example is shown in Figure 2.11, where the figure on the left shows two thin lenses combined in such a way as to produce an image of a distant object on a cryogenically-cooled detector. The figure on the right shows how such a system might look in practice, where the thin lens on the left has been split into two elements, and the lens on the right into three. This is necessary because of WFE known as *aberrations*, and it is the goal of the lens designer to reduce aberrations to the point where image quality is acceptable.

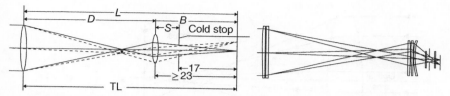

FIGURE 2.11 Comparison of (left) the first-order design of a MWIR imager and (right) a detailed design corrected for aberrations. Adapted from Warren J. Smith, *Modern Optical Engineering*, McGraw-Hill (2008).

Understanding what lens designers do gives us better insight into the challenges of optomechanical design. In this case, the nonlinearity of Snell's law illustrates the key physical dependence of aberrations on local angles of incidence. That is, larger refraction angles have more aberration, as should be expected for rays at steep angles (larger field-of-view or FOV) and far from the optical axis (larger aperture). The first step in reducing aberrations is thus reducing the *first-order* lens powers as much as possible, given the dependence of refractive power on the curvature of the optical surfaces. As this step is often insufficient, the lens designer must also use well-known methods—lens splitting, shape factor, orientation factor, achromatic doublets, and so on—in conjunction with aperture and field control.

One aberration that is common for larger apertures is known as *spherical aberration*. This aberration is observed when a lens or mirror with spherical surfaces images an on-axis point; not surprisingly, it is known as spherical aberration (aka SA, SA3, or simply "spherical"). It cannot be avoided except by using surfaces that are not spherical, and it occurs to varying degrees depending on the aperture diameter. The image quality of fast systems with a small relative aperture ($f/\# = f/D$) is often limited by the lens designer's ability to minimize spherical [2].

Spherical aberration is illustrated in Figure 2.12, which shows an on-axis point at infinity—a star, for example—being focused by a biconvex lens. The figure illustrates that the outer rim ("zone") of the lens brings the wavefronts to a different focus than the central zone near the optical axis. This occurs because the angle between the incident rays and the surface normal of the lens is larger for the outer rays than it is for the central (paraxial) rays near the optical axis. Snell's law shows that the outer rays must bend more; as a result, they are brought to focus closer to the lens while the inner rays are focused farther away. The continuum of zones across the lens thus creates an image blur larger than the perfect geometrical point predicted by first-order optics. This blurring also occurs for other common aberrations—astigmatism and coma for off-axis points, for example—the details of which are reviewed in Kasunic [1] and Smith [2].

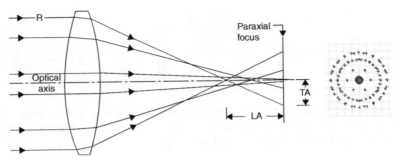

FIGURE 2.12 Spherical aberration results when the outer (marginal) rays are refracted more than the central (paraxial) rays. The difference in foci can be measured as either a longitudinal aberration (LA) or a transverse aberration (TA). Adapted from Warren J. Smith, *Modern Optical Engineering*, McGraw-Hill (2008).

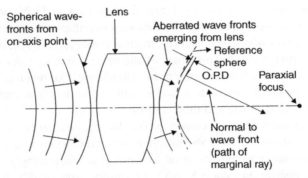

FIGURE 2.13 Aberrations are a result of wavefront error. Adapted from Warren J. Smith, *Modern Optical Engineering*, McGraw-Hill (2008).

While first-order imaging depends on how an ideal lens modifies wavefronts, aberrations are a result of WFE. While the physics is the same as the ray diagram in Figure 2.12, Figure 2.13 shows that spherical aberration can also be viewed as a deviation of the wavefront from the ideal wavefront centered on the paraxial focus (the *reference sphere*). This optical path difference (OPD)—or wavefront error—brings the outer portion of the wavefront to a focus slightly closer to the lens than the inner (or *paraxial*) portion.

The OPD or WFE shown in Figure 2.13 is typically measured in units of wavelengths (or "waves"). For aberrations, it is a maximum at the outer diameter of the lens, so a smaller lens aperture reduces the error. Because of their dependence on incident angle, lens design aberrations can be small for small apertures (large f/#), but also for a small FOV (i.e., the angular size of the object, measured in degrees or radians). If the aperture is small enough, for

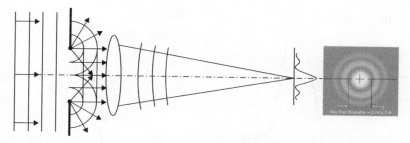

FIGURE 2.14 Diffraction from an aperture results in each point on the object being imaged as a blur whose size depends on the wavelength λ, focal length f, and aperture diameter D. Adapted from Keith J. Kasunic, *Optical Systems Engineering*, McGraw-Hill (2011).

example, the aberrations may be sufficiently small that the image quality may be acceptable. As the aberrations are reduced even further, however, the blur size reaches a point where it is limited not by aberrations but rather by an optical wave property known as *diffraction*.

Ray propagation through an optical system implies that light travels straight through an aperture. Although most of the light does go straight through the aperture, there is also some spreading for the part of the wavefront that reradiates from the aperture's edge as a spherical wave. This spreading is called diffraction, and as shown in Figure 2.14, it results in a point on the object being imaged by a circular lens into a series of concentric bright and dark rings.

Just as with aberrations, the blur of each point created by diffraction reduces the image quality. For circular lenses, the size of the blur is approximated by the Airy disk, which is the diameter of the first dark ring ("first zero") of the Bessel function J_1 that describes the blur [3]. The Airy disk's angular blur size at the image is $\beta = 2.44\lambda/D$, which when projected over the image distance—for example, the focal length for an object at infinity—gives a physical blur $B = 2.44\lambda f/D = 2.44\lambda \times f/\#$. For example, a digital SLR camera imaging a distant scene at visible wavelengths ($\lambda \approx 0.5\,\mu m$) with an $f/2$ lens gives a diffraction-limited blur $B = 2.44\lambda \times f/\# = 2.44(0.5\,\mu m)(2) = 2.44\,\mu m$. The small wavelengths and blur sizes of optical systems are the key reasons for the difficulty of optomechanical design.

The criterion for diffraction-limited imaging was established by John Strutt (Lord Rayleigh) using an optic with extremely small WFE to resolve two point sources such as stars. He found that points are resolved when the diffraction peak of one Airy disk sits on the first dark circle of the other, giving a spacing of $1.22\lambda \times f/\#$ (Fig. 2.15). In addition, it turns out that a peak-to-valley aberration WFE of one-quarter of a wavelength (i.e., $\lambda/4$, or "quarter wave") produces blurs that are only slightly larger than diffraction limited and

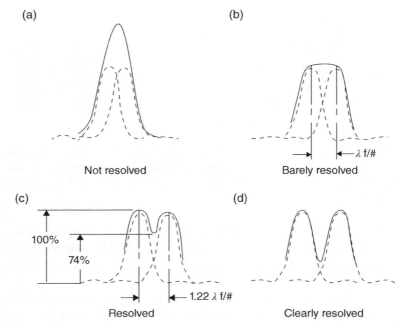

FIGURE 2.15 The overlap of Airy disks from two adjacent point sources determines the best attainable optical resolution for diffraction-limited systems. The commonly used Rayleigh criterion is illustrated in (c). Adapted from Warren J. Smith, *Modern Optical Engineering*, McGraw-Hill (2008).

with a resolution equivalent to the Rayleigh criterion. This is significant because it tells us that *obtaining a diffraction-limited lens does not require zero WFE*; the WFE of an optical system is thus a key measure of resolving power and image quality. For this reason, so-called quarter-wave optics is considered to be the minimum standard for high-quality imagery and is also known as diffraction limited.

WFE can be specified as either peak-to-valley (PV) or root mean square (RMS). For example, California geography with Death Valley (−282 ft) and Mt. Whitney elevations (+14,494 ft) gives a PV surface deviation of 14,776 ft (4.4 km); the RMS deviation averaged over all of California is much smaller. By removing outliers such as 6-sigma (6σ) deviations, the RMS WFE—that is, the standard deviation from the average WFE at 100 or so points across the lens aperture—is a better statistical measure of image quality. As a result, vendors typically specify optics using RMS wavefront error; see Figure 2.16 and Chapter 3.

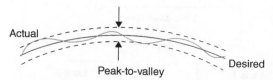

FIGURE 2.16 Peak-to-valley (PV) wavefront error is typically larger than RMS WFE by a factor of 5× for random fabrication errors, but a defocus factor of 3.5× is also used.

The relationship between PV and RMS WFE depends on the type of aberration. Defocus is often used for WFE budget comparisons between RMS and PV such that $W_{PV} = 3.5W_{RMS}$, for which a diffraction-limited system corresponds to $\lambda/(4 \times 3.5) = \lambda/14$ RMS or ~0.07 waves RMS. Note that care must also be taken with the wavelength of measurement versus the wavelength of use: 0.07λ RMS is diffraction limited, but what is λ?

It is necessary for individual elements within a complex lens to have WFE $< \lambda/4$ PV for the optical system to maintain diffraction-limited performance, with each element typically on the order of $\lambda/10$ PV or less. One reason is that the WFE for multiple elements must add to a total system WFE $\approx \lambda/4$. Another reason is that optomechanical factors contribute to the overall WFE, including lens-to-lens alignments and fabrication errors such as imperfections in each lens's radius of curvature. Fabrication and alignment issues are examined in Chapters 3 and 4, respectively. Example 2.2 shows how to use the results from those chapters to create a WFE budget, a commonly used optical engineering tool for assessing the relative contributions to diffraction-limited performance.

Example 2.2   The total WFE for a lens assembly is required to be diffraction limited, with the WFE $= 0.071\lambda$ ($= \lambda/14 = 0.633\,\mu m/14 = 0.045\,\mu m$) RMS at $\lambda = 633$ nm. The assembly incorporates two lenses, both similar in design and fabrication; we therefore initially assign ("allocate") a WFE $\approx 0.05\lambda$ for each lens so that the root-sum-of-squares (RSS) of the lenses adds to a total WFE of 0.071 waves for the assembly.

Figure 2.17 shows how the WFE budget for each lens flows down to the components that contribute to the WFE for each lens. Included are WFE entries for lens design, fabrication, initial alignment, and temperature and vibration-induced misalignments from the operating environment. The *lens design residual* is the WFE due to remaining aberrations; this is not necessarily the smallest WFE the lens designers can deliver but only the smallest necessary to meet the $\lambda/14$ spec for the overall assembly. Note that the design WFE for each lens must be significantly smaller than diffraction limited to meet this spec.

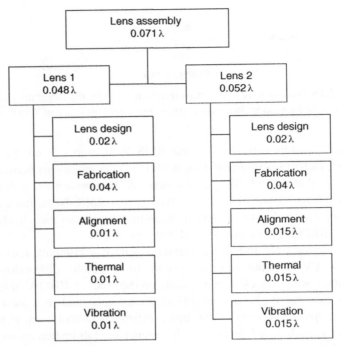

FIGURE 2.17 Allocation of wavefront error budget for a two-lens optical assembly (e.g., an objective and a field flattener) that includes the effects of design, fabrication, alignment, thermal changes, and vibration.

The other four contributors to WFE are optomechanical terms and will be covered throughout the rest of this book. The five components shown in Figure 2.17 are independent of each other, so they add as an RSS to yield the total WFE for lens 1: $WFE_1 = [(0.02)^2 + (0.04)^2 + (0.01)^2 + (0.01)^2 + (0.01)^2]^{1/2} = 0.048$ waves RMS; a similar sum is computed for lens 2, giving 0.052 waves RMS and a system $WFE = (WFE_1^2 + WFE_2^2)^{1/2} = [(0.048)^2 + (0.052)^2]^{1/2} = 0.071$ waves for the assembly. The WFE budget is useful because it shows the term that dominates the total error; in this example, it is the lens fabrication with $WFE = 0.04\lambda$. Therefore, designer time would best be spent on reducing this item if the lens assembly is intended to be used with additional optics. Alternatively, the small values allocated for the alignment, thermal, and vibration terms may require an extremely difficult alignment procedure or optomechanical design to obtain, and may be traded off (increased) against possible reductions in the fabrication term to reduce the risk of failure.

Summary: While lens-design aberrations create WFE, there are many other sources of WFE, including fabrication errors, initial alignment errors, and environmental factors such as external vibrations, temperature changes, temperature gradients, and so on. The aberrations may be reduced by the lens designer; the fabrication, alignment, and other wavefront errors are the focus of the optomechanical design and are reviewed in the remainder of this book under the categories of fabrication (Chapter. 3), alignment (Chapters 4 and 9), structural design (Chapters 5, 6, and 7), and thermal design (Chapter 8).

PROBLEMS

2.1 Using a mechanical analogy for Snell's law, which way does a car turn if it starts out in the sand and drives onto the asphalt road? Is there an angle the car can drive onto the asphalt where it doesn't turn?

2.2 Physically, why does a larger refractive index n in Equation 2.1 result in a lens with a shorter focal length?

2.3 Using Equation 2.1, what is the focal length of a flat window with $R_1 = R_2 = \infty$?

2.4 The refraction at a surface can be defined as the difference between the incident angle θ_i and the transmitted angle θ_t (both measured with respect to the surface normal). Show that $\theta_i - \theta_t$ depends on the incident angle and the ratio of refractive indices n_i/n_t. Is your equation physically reasonable? That is, does the refraction increase with θ_i and n_t/n_i?

2.5 How would you obtain the lenses with extremely small WFE used by Lord Rayleigh? Is any special lens-design method needed or can a high-quality off-the-shelf lens be used?

2.6 Would you expect the tilting misalignment of a lens with a short focal length to have more or less WFE than a lens with a long focal length? Hint: How do the incidence angles on the lens surfaces change with focal length?

2.7 If the fabrication error for both lenses in Example 2.2 could be reduced to 0.02λ, how much could the alignment errors be increased and still maintain diffraction-limited performance for the system?

2.8 In Example 2.2, a wavelength of $0.633\,\mu m$ was used to design and measure the lens assembly. Why? What is the WFE (in units of waves) of the assembly if we use it at a wavelength $\lambda = 1.266\,\mu m$?

REFERENCES

1. K. J. Kasunic, *Optical Systems Engineering*, New York: McGraw-Hill (2011).
2. W. J. Smith, *Modern Optical Engineering* (4th Edition), London: McGraw-Hill (2008).
3. E. Hecht, *Optics* (4th Edition), Reading: Addison-Wesley (2001).

3

OPTICAL FABRICATION

On the walk from the parking lot into the building where you work, you may have noticed that the reflection from the windows does not produce a particularly good image of the scene behind you. This is especially prevalent if there are multiple windows next to each other, creating a huge wall of glass that reflects the entire scene. The scene in each pane is almost always distorted— and yet, when you get inside the building and look through those same windows, the transmitted scene looks perfectly acceptable. While mirrors will in general produce more image distortion than a window, the lack of distortion looking through the window is also a result of the fabrication process for thin optics, where the surface errors that produce poor images in reflection are compensated by those on the other side of the window (Fig. 3.1).

Other fabrication parameters also affect optical performance, including variations in the refractive index, surface roughness, and dimensional control. For example, Figure 3.2 shows a high-performance segmented window, whose panes each have a small amount of fabrication error known as *wedge*. This angular error changes the image location slightly, and thus the wavefronts transmitted through each pane may not overlap on the focal plane with sufficient fidelity to produce a good image. In this case, then, a small fabrication error in one of the simplest of optical components—a flat window—can produce significant loss of performance.

Optomechanical Systems Engineering, First Edition. Keith J. Kasunic.
© 2015 John Wiley & Sons, Inc. Published 2015 by John Wiley & Sons, Inc.

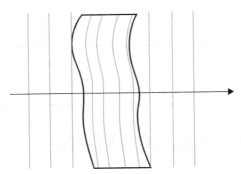

FIGURE 3.1 With correlated surface errors—as sometimes found with a thin window—the wavefront errors induced by the first surface are compensated by those of the second for on-axis wavefronts. Credit: Ray Williamson, *Field Guide to Optical Fabrication*, SPIE Press (2011).

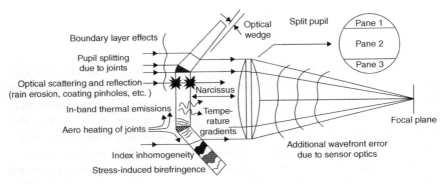

FIGURE 3.2 The wedge in each pane of a high-performance segmented window can dominate the image quality. Credit: Michael I. Jones and M. S. Jones, Proc. SPIE, Vol. 1498 (1991).

The fabrication of standard-size optics (10–100 mm in diameter) begins with the rough shaping of a block of glass (BK7, SF11, etc.) or crystalline material (silicon, germanium, etc.). After machining ("generating") the surfaces roughly to shape, the resulting blank is mounted ("blocked") with other blanks for grinding with finer abrasives in preparation for the final polishing steps. Along with material quality, this polishing is a critical factor in determining the image quality produced by the lens.

Lenses can be polished to spherical, cylindrical, or aspheric shapes. Beamsplitters, windows, prisms, filters, and polarization components usually have flat surfaces. Mirrors can be flat, spherical, or aspheric—including parabolic, elliptical, hyperbolic, or polynomial. The focus of this chapter is on the final polishing of standard-size optics with spherical and flat surfaces. The asymmetry of aspheres makes them difficult to polish, although diamond turning of such surfaces is common for mid-wave infrared and long-wave infrared (LWIR) materials. For further details on the fabrication and test of aspheres, see Williamson [1], Anderson and Burge [2], Goodwin and Wyant [3], Malacara [4, 5], and Karow [6].

Optical fabrication requires a prescription provided by the lens designer (Table 3.1), from which optomechanical drawings are created for each component (Fig. 3.3). These drawings are then used by an optical fabrication vendor (or "shop") to grind and polish the lens into the shape specified on the drawing. Fabrication quality is measured by a number of metrics. For imagers, wavefront error (WFE) is one of the most important, and a poorly fabricated lens will degrade the image resolution to much worse than the diffraction limit. A fabrication metric often used to quantify these errors is surface figure

TABLE 3.1 Component Prescription for the Lens Shown in Figure 3.4

Surface	Radius	Thickness	Glass	Semi-diameter
1	13.589	4.45	BK7	10.3
2	64.643			14.5

All dimensions are in units of millimeters.

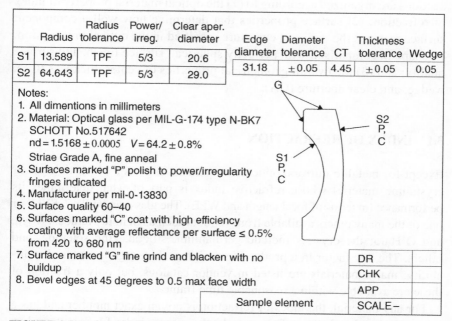

FIGURE 3.3 Conventional component drawing used by optical vendors to fabricate lenses. Drawings using the ISO 10110 standard are also common [7]. Adapted from R. E. Fischer, B. Tadic-Galeb, and P. R. Yoder Jr., *Optical System Design*, McGraw-Hill (2008).

error (SFE), and we will we see in this chapter how to relate this to metric to the resulting WFE for windows, lenses, and mirrors.

The component drawings can be surprisingly complex as they must specify every detail that is important, including the tolerances. Optical tolerances are calculated by the lens designer, whose responsibility it is to find a design that does not entail tight fabrication and alignment tolerances. The tolerances must be determined in conjunction with the fabrication engineer, optomechanical engineer, thermal analyst, and structural analyst to determine whether or not the system can even be built (i.e., is "producible"). Tolerances are typically classified as standard ("loose"), precision, and high precision ("tight"). Volume quantities of low-cost, commercial off-the-shelf lenses are fabricated to standard tolerances; at the other extreme, "one-off" state-of-the-art designs usually involve high-precision (and thus high-cost) tolerances. Looser tolerances are generally more producible up to the manufacturer's standard tolerance level, at which point looser tolerances are no longer useful. Even if we are simply buying off-the-shelf optics, however, it is important to understand the various tolerance and quality levels, so we do not spend more money than we need to on a lens that has better performance than is necessary [8].

The fabrication parameters and tolerances reviewed in the following sections are organized according to (1) the optical material property of index of refraction; (2) surface properties that determine how well a component surface matches the radius of curvature specified in the optical prescription, its shape in comparison with perfectly spherical, surface finish, and surface quality; and (3) the overall geometrical properties of center thickness (CT), wedge, and clear aperture (CA).

3.1 INDEX OF REFRACTION

Except for metallic mirrors, optical components start as a block of glass or crystalline material whose refractive index is part of what determines its performance (in terms of focal length and WFE). The glass can be BK7, SF11, or one of the many others available from vendors such as Schott, Heraeus, Hoya, and O'Hara; the crystals include germanium, silicon, zinc selenide, and others. The key factor in a producible design at this stage is material availability; many materials are listed in vendor catalogs, but only a fraction of these are available within a reasonable time frame.

For any material, the index of refraction is not an exact number and has a tolerance associated with it. For example, the average index for Schott product number 517642 at a wavelength of $\lambda = 587.6$ nm is $n_d = 1.5168 \pm 0.001$ for standard-grade glass. Table 3.2 shows that higher-precision grades are also

TABLE 3.2 Standard, Precision, and High-precision Tolerances Associated with Various Index-of-Refraction Properties for Glasses[a] [2]

Property	Standard	Precision	High precision
Index of refraction	±0.001	±0.005 (Grade 3)	±0.0002 (Grade 1)
Homogeneity	±10^{-4}	±5×10^{-6} (H2)	±10^{-6} (H4)
V-number	±0.8%	±0.5% (Grade 3)	±0.2% (Grade 1)
Birefringence	10 nm/cm	6 nm/cm	4 nm/cm
Striae	30 nm (Grade C)	15 nm (Grade B)	<15 nm
Bubbles/inclusions	0.5 mm^2 (Class B3)	0.1 mm^2 (Class B1)	0.03 mm^2 (Class B0)

For high-index glasses (i.e., with $n_d > 1.83$), the tolerances on refractive index are doubled. Material property data for the correct wavelength and temperature range is often difficult to obtain, so approximations must be used. Vendor catalogs are excellent sources of data, including Schott' s Web site for glass data (www.schott.com) and Crystran's publication [9].

[a] Adapted from Keith J. Kasunic, *Optical Systems Engineering*, McGraw-Hill (2011).

available for glasses, with tolerances that are 2× and 5× smaller than the standard grade.

In addition to the tolerance on average index, there are also small index variations throughout a volume of optical material. Known as index inhomogeneities, these spatial variations result in WFE across a lens that are similar to the atmospheric distortions that cause shimmer. Variations throughout the volume of large blocks of glass that come out of a melt furnace are $\delta n \approx \pm 10^{-4}$. Measured in radians, the resulting wavefront phase error is $2\pi\delta n(t/\lambda)$ radians; to obtain the WFE in cycles or waves, this value is divided by 2π radians per cycle; thus,

$$\text{WFE} = \delta n \frac{t}{\lambda} \quad \text{[waves PV]} \tag{3.1}$$

where PV denotes "peak-to-valley" or maximum-to-minimum. At a wavelength of 0.5 μm, an average index variation $\delta n = 2 \times 10^{-4}$ (PV), and an element thickness $t = 10$ mm, the resulting WFE = 4 waves (4λ) PV, a very large fabrication error for precision optics designed to $\lambda/4$ PV or better. Fortunately, the inhomogeneity of smaller pieces of glass that are cut out of the melt to fabricate standard-size optics are on the order of 10^{-5} to 10^{-6}, which results in a WFE that is smaller by one or two orders of magnitude. In addition, the conversion of the PV WFE to root-mean-square (RMS) WFE entails an averaging over the area of the optic (see Example 3.1), reducing the WFE by yet another factor that depends on the inhomogeneity distribution.

Birefringence ("double refraction") occurs naturally in some materials; in others, it is seen when the material is stressed. It is observed as a focal length that is different for each of the two principal polarizations, a result of their different refractive indices (the "ordinary" and "extraordinary"). Many materials

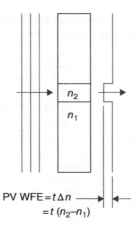

$$PV\ WFE = t\Delta n$$
$$= t\ (n_2 - n_1)$$

FIGURE 3.4 Effects of high-index striae on the wavefront error for an optic with a high-precision striae spec. The index difference may not extend through the entire optic, in which case the WFE is smaller. Adapted from Keith J. Kasunic, *Optical Systems Engineering*, McGraw-Hill (2011).

are not birefringent; however, calcite and crystalline quartz have inherent birefringence and so are not usually used except for polarization components. Plastics are highly birefringent when stressed, and a few materials (e.g., SF57) have a small stress-induced birefringence.

Stress can result from external forces, but there are also internal stresses that are "locked in" to the material during the cooling process in manufacture. It is these internal stresses that are specified to the material supplier or fabrication facility. The spec is given as the stress-induced index difference δn_b, which is measured in units of nm/cm. A typical value for δn_b is 10^{-6}, which can be written as $(10 \times 10^{-9}\,m)/(10^{-2}\,m) = 10\,nm$ of phase difference per cm of glass thickness (units of nm/cm); this is an appropriate spec for visual instruments such as photographic and microscopic instruments. The highest precision is used for components such as polarizers and interferometers, which require very low birefringence. Glasses with $\delta n_b \approx 4\,nm/cm$ are available that have been precision annealed to remove the locked-in stresses for such applications.

Another index-of-refraction property known as *striae* must also be specified. Striae (striations) are small localized regions of index inhomogeneities that may be caused by temperature gradients that occur as the molten glass from the melt furnace cools off. There are various grades for classifying striae, depending on their visibility and orientation with respect to the surface of the element; striae that extend throughout the direction of propagation impart the most WFE (Fig. 3.4). For a material that meets the high-precision grade, the effects on WFE are small: parallel striae values of $\delta n = 10^{-6}$ produce a

WFE $= t\,\delta n = 0.01$ m $\times 10^{-6} = 10$ nm ($\lambda/50$ at $\lambda = 0.5$ μm) in a 10-mm-thick optic. Thinner elements have proportionally less WFE, which may allow for the use of a less expensive, coarser-grade striae spec.

Finally, optical materials can also have highly localized regions, such as air bubbles and inclusions, with large index differences that cause reflections and scattering. These can sometimes be seen as faint shadows when found in optics such as field flatteners that are located near an image. Table 3.2 lists the acceptable size of these imperfections for various material classes, where it is understood that the total allowable area of the bubbles and inclusions (in units of mm^2) is per 100 cm^3 of material volume. Only smaller imperfections are acceptable for applications requiring very low scatter and for high-power laser systems in which absorption at metallic inclusions creates heating and fracture of the optic.

3.2 SURFACE CURVATURE

Starting with known refractive-index properties, a material can be fabricated into lenses, prisms, and other components. The process entails grinding and polishing the element surfaces with the flat, spherical radius, or aspheric shape given in the optical prescription. As we have seen in Chapter 2 for spherical surfaces, the index and curvature determine the refractive power of each surface; for aspheric surfaces, the surface power is not constant and varies across the aperture. In either case, a surface cannot be polished into a perfect size or shape; for spherical surfaces, there are tolerances associated with the radius that affect the optic's refractive (or reflective) power and hence its focal length (Fig. 3.5).

The traditional methods for tolerancing the power of spherical surfaces are with spherometers or by comparison with a precision reference optic known as a test plate; more recently, interferometers are also used to make these measurements [3]. The transparent test plates are fabricated to very high precision with opposite curvature—for example, a concave test plate is used to measure a convex surface—so that a correctly fabricated element fits into the test plate with only a small error (Fig. 3.6). The test-plate fit (TPF) thus measures how closely the radius of curvature of the fabricated surface matches the design ideal when the test-plate radius was used to create the optical prescription.

The error in curvature is measured as the gap between the surface being polished and the test plate. For spherical surfaces, it is observed as contour fringes (Newton's rings) when looking at how well the surface fits the test plate (Fig. 3.7); in reflection, a fringe—that is, one bright plus one dark ring—represents a distance of one-half of a wavelength (PV) between the surface and the test plate [3]. For example, an incorrectly polished surface may be

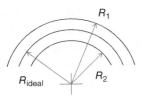

FIGURE 3.5 The ideal radius R of a spherical lens or mirror surface has a tolerance ΔR associated with its fabrication.

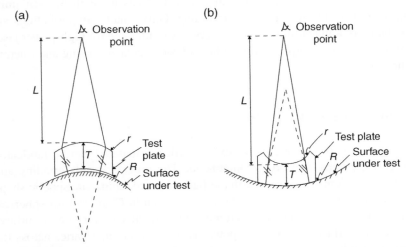

FIGURE 3.6 The curvature of a convex (a) or concave (b) optical surface is commonly measured by comparison with a known reference called a test plate. Credit: Daniel Malacara, *Optical Shop Testing* (3rd Edition), John Wiley & Sons (2007).

perfectly spherical yet still deviate by a surface error $\Delta R = 1\,\mu m$ (for instance) from the required curvature; this deviation is measured by the test plate as $N = 2\Delta R/\lambda = 4$ circular fringes at a wavelength of $0.5\,\mu m$. These are observed as four bright and four dark rings—or lines for a flat optic—across the aperture in an alternating sequence indicative of the constructive and destructive interference that occurs as the optical path difference (OPD) between the surfaces increases or decreases in half-wavelength increments.

The wavelength of the measurement must therefore be specified, since a TPF that produces four fringes at a wavelength of $0.5\,\mu m$ produces only two fringes at a test wavelength of $1\,\mu m$. A typical test wavelength for optics that transmit in the visible is 632.8 nm, the red color of the helium–neon (HeNe) laser. Thus, a lens fabrication spec for each surface may read as follows: "Test plate fit over the clear aperture shall be no more than four fringes of power at a wavelength of 633 nm." It is also necessary to relate the measurement to the

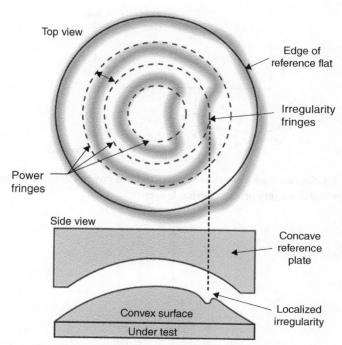

Top view

Edge of
reference flat

Irregularity
fringes

Power
fringes

Side view

Concave
reference
plate

Convex surface

Under test

Localized
irregularity

FIGURE 3.7 Interference pattern produced by an imperfect spherical surface when measured with a test plate. Courtesy of Edmund Optics, Inc.

design wavelength because ultimately it is the optical system performance that is important.

A precise mapping of the surface curvature (or "power") to the test plate is often not critical; typical specs for the refractive power of a surface are on the order of 4–10 fringes. For example, the column titled "Radius tolerance" in Figure 3.3 refers the fabrication shop to the TPF, which is listed in the "Power/Irreg." column as 5 fringes of power. This may seem to be a large number of fringes for an optic with a diffraction-limited $\lambda/20$ WFE spec (for example); as shown in Section 3.3, the smaller errors responsible for most component WFE are characterized using what is known as a surface figure (or "irregularity") spec.

3.3 SURFACE FIGURE

The tolerance on surface radius determines how closely a spherical or flat surface fits that of a test plate; another property known as the *surface figure* specifies the magnitude of small-scale surface irregularities such as those shown in

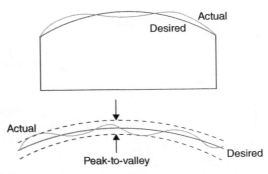

FIGURE 3.8 Surface figure errors are an unavoidable product of the fabrication process, with a higher-quality process having less error.

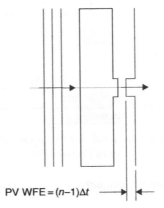

PV WFE = $(n-1)\Delta t$

FIGURE 3.9 Wavefront error created by surface irregularities in a flat plate. The surface figure error (SFE) is the height (or depth) Δt of the irregularity, the resulting PV WFE $= (n-1)\Delta t$ for an optic in air.

Figure 3.8. These irregularities from the polishing process may be found in the fabrication of any surface, including flats, cylinders, spheres, and aspheres. Deviations from the ideal surface induce WFE (Fig. 3.9) and associated aberrations such as coma or astigmatism, which reduce image quality. The surface figure WFE is in addition to the effects of unavoidable design residuals and refractive-index variations.

TPFs, interferometers, and profilometers are used to evaluate surface figure, where the deviation (or "irregularity") from circular fringes measures the variation from a spherical surface; this is shown in Figure 3.7 as fringes of deviation from the circular rings defining power. Note that it is difficult to measure irregularities using test plates that are a small fraction of the power spec, and a ratio of 4–5 fringes of power for every fringe of irregularity is a common measurement limitation for production optics. Interferometers have no such a limitation and are used for high-precision optics. Independent of the fringe ratio, the resulting WFE is $WFE = (n-1)\Delta t/\lambda$ due to a PV $SFE = \Delta t/\lambda$ in units of wavelengths; in terms of test-plate fringes, the PV WFE for each surface is given by [10]

$$WFE = (n-1)\,SFE = \frac{1}{2}(n-1)N \quad [\text{waves PV}] \qquad (3.2)$$

for a surface measured in air with index $n_a = 1$; the factor of 1/2 is a result of using test plates, as they double the OPD due to surface mismatch. For $N = 1$ fringe of irregularity and a material with index $n = 1.5$, this corresponds to 0.25 waves ($\lambda/4$) of PV WFE—equal to that required for a diffraction-limited lens. A lens has two surfaces, so a spec on surface figure for a diffraction-limited lens in the visible might read as follows: "Test-plate fit over the clear aperture shall be no more than 1/2 fringe of irregularity at a wavelength of 633 nm."

Unlike those shown in Figure 3.1, the WFE errors on the surfaces of lenses are usually generated randomly with respect to each other; hence the errors add as a (square) root sum-of-squares (RSS) for many surfaces (and refractive indices) in a system. For a system consisting of a single lens with uncorrelated fabrication errors, the WFE of the lens is

$$WFE = \sqrt{[(n-1)\,SFE_1]^2 + [(n-1)\,SFE_2]^2 + (\delta n \cdot t)^2} \quad [\mu m] \qquad (3.3)$$

Specifying one-half fringe of irregularity then seems a bit conservative for volume manufacturing, given that 0.707 fringes for each of two surfaces will RSS to produce 1 fringe of WFE. However, this fails to take several factors into account: the refractive-index variations δn, the design contribution to WFE, the possible existence of multiple elements in a lens assembly, and the alignment and environmental effects reviewed in the following chapters. It is thus common to use irregularity specs as low as 0.05 waves ($\lambda/20$) for high-precision applications, such as interferometers or photolithography optics, and to use elements with a small-enough aspect ratio (e.g., $\leq 6:1$ ratio of diameter and thickness) that they will not be deformed during polishing and coating to such tight tolerances.

The fabrication of mirrors with low WFE is particularly difficult because a reflected wavefront interacts twice with surface figure error. This phenomenon is analyzed by using a refractive index $n = -1$ in Equation 3.2 for a mirror in air; this gives a $WFE = 1$ wave for 1 fringe of irregularity, larger by a factor of 4 than that for a refractive lens surface with $n = 1.5$. The aspect ratio is thus critical for mirrors, as is using the correct substrate material. Zerodur is stiff enough for a $\lambda/20$ irregularity spec, but N-BK7 and fused silica are not[1]—see Chapter 6.

Wavelength also plays a major role in the smallest WFE that can be obtained. A lens with 1 fringe of irregularity in the visible, for example, has a surface figure error on the order of $0.25\,\mu m$, and the surface must be accurately polished to within this value. In contrast, an LWIR lens with the same refractive index at a wavelength of $10\,\mu m$ can have an SFE approximately of $5\,\mu m$ (20 times larger) before the WFE reaches 1 fringe, a much easier tolerance to maintain during fabrication.

Finally, an important distinction when fabricating components is that between the PV error shown in Figure 3.8 and the RMS SFE. That is, a lens or mirror may have a surface error that is randomly distributed over the component area; alternatively, the surface error may be localized over a small area. If the SFE is localized—as is the irregularity in Figure 3.7—the effects on image quality can be much smaller than if the WFE is random. The use of an area-averaged metric—that is, RMS SFE—thus provides a better measure of component fabrication quality. Example 3.1 illustrates the importance of the RMS concept.

Example 3.1 Due to inadvertent finger grease, the center window in Figure 3.2 has a large surface error; the error is approximately 5λ PV over the area of the grease ($1\,mm^2$). What are the effects of the grease on the WFE of the system? The window is $100\,mm \times 100\,mm$ in size; assume its PV SFE is ≈ 0.

If the surface errors were random, then the RMS WFE would be approximately $5\times$ smaller than the PV WFE [10]. Finger grease is not a random fabrication error, however, and the effects on WFE can be much smaller. For optics that are at or near the collection aperture (or *entrance pupil*)—such as a window located before a collection lens—the RMS WFE is the better measure of image quality, as the wavefront samples the entire aperture, and thus averages the PV WFE over the area. If the area of the surface error is small, $WFE_{RMS} \approx WFE_{PV} \times (A_{PV}/A_{optic})^{1/2}$, where A_{PV} is the area of the PV surface error ($1\,mm^2$ in this example) and A_{optic} is the much-larger area of the window, or $100\,mm \times 100\,mm = 10^4\,mm^2$ in this example.[2] The RMS WFE from the finger grease is thus $WFE_{RMS} \approx 5\lambda \times (1/10^4)^{1/2} = 0.05\lambda$. This is about the same as the RMS WFE that would be allowed—approximately $\lambda/20$—for a diffraction-limited system with random

---

[1] As a rule of thumb, a minimum thickness of 2 mm is required for flatness during fabrication of standard-size elements, independent of aspect ratio.

[2] Quantitatively, the conversion from PV to RMS depends on the area weighting of the PV errors, such that $SFE_{RMS} = \left[\sum w_i SFE_i^2\right]^{1/2}$, where SFE_i is the PV surface error for each area element with area A_i, and $w_i = A_i/A_t$ for a total area of the optic A_t. In addition, the SFE_i must first be adjusted to give an area-weighted mean $\overline{SFE} = 0$ (see Fig. 3.11 for a surface roughness RMS calculation). If $A_i \times SFE_i^2$ for the finger grease is much larger than that of the rest of the optic, the irregularity dominates the RMS SFE and the equation given in the text is obtained.

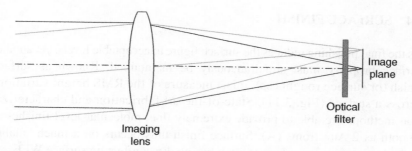

FIGURE 3.10 The sampling of surface errors by the wavefront does not always capture the entire size of the optic, reducing the induced wavefront error.

fabrication errors and is the reason that a small, thin fingerprint on a large camera lens does not cause noticeable drops in image quality.

If the optical component is not located at a pupil, but instead near an image, then the scaling of PV WFE can be a bit more involved. Figure 3.10 shows an optical filter located near an image plane—do we need to specify high-precision, diffraction-limited tolerances on the fabrication of this component? For random fabrication errors, it would seem that we need to specify better than $\lambda/4$ PV optical quality ($\lambda/20$ RMS) over the entire aperture, given the RSS addition of other sources of WFE in the system (the imaging lens, for example). Fortunately, only those lenses that are at or near a pupil (entrance, exit, or intermediate) must meet the irregularity spec over the full diameter of the lens.

Figure 3.10 shows why. The figure shows that the on-axis wavefront incident on the imaging lens covers its entire diameter, while the focused wavefront incident on the filter does not. The focused wavefront thus samples a smaller fraction of the surface-figure errors than the unfocused wavefront incident on the imaging lens. The diameter of the smaller cone of light incident on the optic thus has fewer fringes of irregularity than the larger filter size over which the surface figure is often specified—in proportion to D_{cone}/D_{optic} for random fabrication errors equally distributed over the aperture. The conversion from PV WFE to RMS then proceeds in the usual way, with $\text{WFE}_{RMS} = \text{WFE}_{PV}/5$ [10].

As a result, commercial optics can perform better than expected when used away from a pupil, or the cost of fabricated optics can be significantly reduced by specifying the irregularity only over the required cone size (approximately 10 mm for a 5× afocal system with a 50-mm entrance pupil, for example). The spec for a 50-mm diameter optic might therefore read as follows: "Test-plate fit shall be no more than 1/2 fringe of irregularity at a wavelength of 633 nm over any 10-mm diameter over the clear aperture." (Here 10 mm has been used to illustrate the concept, not as an example of how all component specs should be written.) That being said, it is also important not to place a refractive optic such as a filter or field flattener too close to an image, as scratches and surface flaws can then be projected onto the image plane—see Section 3.5 for details.

3.4   SURFACE FINISH

As the final polishing reduces the surface figure to acceptable levels, yet another surface property must simultaneously be maintained. Known as surface finish (or surface roughness), it is a measure of the RMS height variations across a surface (Fig. 3.11). State-of-the-art fabrication and characterization methods are able to provide extremely fine, molecular-level finishes as smooth as 2 Angstroms (Å). Surface finish thus occurs on a much smaller scale than surface irregularities, so it has no direct effect on surface WFE.

However, poor surface finish increases the scattering of a surface, which is measured as angular scattering (bidirectional scatter distribution function) or as the sum over all angles (total integrated scattering). Most applications do not require a super-polished surface with only 2-Å microroughness; more typical specs that are uniquely determined by the stray-light requirements for each system are 10–50 Angstroms (Å). Metals such as beryllium and aluminum are "sticky" and typically cannot be polished to better than about 50 Å, so they are coated with electroless nickel when used as mirror substrates requiring better surface finish. In contrast, fused silica and ultra-low expansion glass can be polished to very fine finishes—down to 5 Å if required—but cannot be diamond turned, as the glass breaks from the diamond tool in a brittle manner, resulting in excessive surface roughness.

Along with surface figure, surface finish can be a major cost component because meeting these specs involves a time-consuming process of fine

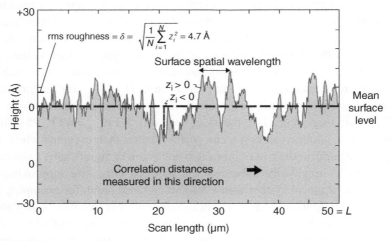

FIGURE 3.11 Surface finish of a polished or diamond-turned optic is measured as an RMS surface roughness. Credit: J. M. Bennett and L. Mattsson, *Introduction to Surface Roughness and Scattering*, Optical Society of America (1989).

polishing. It is expensive to simultaneously obtain high-precision figure and finish since the additional polishing required to give good finish often affects the figure. A recent fabrication method that avoids this trade is magnetorheological finishing (MRF), which uses a magnetic material (iron) in the polishing fluid to control figure and finish. The method is useful for surface figures down to $\lambda/20$ as well as surface finishes to about $10\,\text{Å}$ RMS [11].

3.5 SURFACE QUALITY

The polishing process can leave marks on surfaces that are too large for the finishing process to remove. These are the types of scratches often seen on a windshield or pair of eyeglasses, but they also include larger pits and divots known as "digs." They are mostly cosmetic defects but can be seen as image shadows when found on lens surfaces—for example, field flatteners and field lenses—that are located near the image. They can also be a concern in systems where stray light is a concern, as well as in high-power laser systems where enhanced electric field strength at these small-area imperfections can destroy the optic.

Scratch and dig are specified in terms of allowable sizes, spacings, and densities. In its simplest form, the scratch is measured in units of length and specified as 10 times larger than the scratch width in microns. A scratch specification of 60, for example, corresponds to surface scratches that are no more than $6\,\mu m$ in width.

The dig spec is also in units of length but is given as 10 times smaller than the actual diameter in microns; a dig spec of 40 thus refers to a dig with a 400-μm diameter. Both scratch and dig are usually examined visually by an experienced inspector. Table 3.3 summarizes the tolerances for the surface quality (scratch/dig), as well as the surface properties of surface radius (power), surface figure (irregularity), and surface finish

TABLE 3.3 Standard, Precision, and High-precision Tolerances Associated with Surface Properties[a] [2, 13]

Property	Standard	Precision	High precision
Surface radius (PV)	5 fringes	3 fringes	1 fringe
Surface figure (PV)	2 fringes	0.5 fringe	0.1 fringe
Surface finish (RMS)	50 Å	20 Å	10 Å
Surface quality	80/50	60/40	20/10

[a] Adapted from Keith J. Kasunic, *Optical Systems Engineering*, McGraw-Hill (2011).
Surface radius is specified as "power"; surface figure is specified as "irregularity."

for standard, precision, and high-precision tolerances. Not included are midspatial frequency properties such as "quilting" or "ripple," important surface properties for laser systems and high-precision radiometric instruments [12]. As with the index-of-refraction tolerances in Table 3.2, the cost of high-precision tolerances is significantly higher than that of standard tolerances.

3.6 CENTER THICKNESS

It is possible to fabricate a lens with exactly the correct index and radii as called out in the prescription but still end up with the wrong focal length. This outcome is due to the lens thickness—shown in Equation 2.1 as the variable CT, which is set to zero for thin lenses. For thick lenses, the distance traveled in the lens affects the height at which the rays strike the second surface, changing the focal length and WFE of the optic. Clearly, the steeper the rays and the thicker the optic, the larger are the effects of thickness variations.

The thickness must therefore be toleranced, but the tolerance may refer to the CT, the edge thickness, or somewhere in between. A common standard is the CT (Fig. 3.12), although this is unfortunately a difficult parameter to control. Typical thickness tolerances are on the order of ±150 μm for standard optics in volume production and ±25 μm for high-precision elements. This precision is

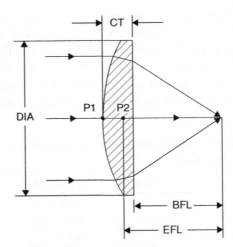

FIGURE 3.12 The center thickness (CT) is commonly used to specify a lens thickness. Also shown are the principal points (P1, P2), effective focal length (EFL), and back focal length (BFL). Reproduced with permission from Janos Technology LLC.

not possible for softer materials, for which the normal variations in grinding pressure allow less control over the material removed during polishing.

3.7 WEDGE

Yet another fabrication tolerance refers to the angular relationship (tilt) between surfaces. A plane-parallel plate such as the window shown in Figure 3.2, for example, is specified as having surfaces that are both flat (e.g., a surface figure of $\lambda/10$) and parallel to some tolerance. If one surface is tilted with respect to the other (Fig. 3.13)—as all surfaces are, to some degree—then rays will deviate from their expected path. It is therefore necessary to quantify the degree to which surfaces must be parallel. For elements that are flat on both sides, this tolerance is known as *wedge* or *surface tilt*; for lenses, it may also be called *centration*.

Looking first at planar elements such as windows, we see that the two sides cannot be polished perfectly parallel; hence, parallel rays that pass through the element will change direction. This is the basis of a thin prism, where the refractive index and wedge angle determine the angular deviation δ [10]:

$$\delta = (n-1)\alpha \quad [\text{rad}] \tag{3.4}$$

The wedge angle α between the two surfaces is specified as a maximum for flats, where typical numbers for standard optics are on the order of 5 arcminute. (An arcminute—sometimes written simply as a "minute"—is a very old unit still used by opticians and astronomers; it is equivalent to 1/60 of a degree, or approximately 291 µrad.) For a material with a refractive index $n = 1.5$, a wedge of 5 min produces a deviation $\delta = (n-1)\alpha = 0.5 \times 1.455$ mrad ≈ 0.73 mrad—enough to produce a linear displacement of $d = \delta \times L = 730$ µm when propagated over a distance $L = 1$ m.

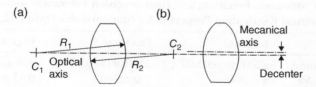

FIGURE 3.13 Lens fabrication tilt is created by an offset ("decentration") between the mechanical axis determined by the edge grinding and the optical axis defined by the centers of curvature of the surfaces. For the lens in (a), the optical and mechanical axes are coincident, while in (b) they are not. Adapted from Keith J. Kasunic, *Optical Systems Engineering*, McGraw-Hill (2011).

For lenses, the optical axis defined by the center of curvature of the surfaces will be offset from the mechanical axis defined by the edge grinding so that the outer diameter (OD) is not concentric with the diameter defined by the optical axis. This is shown in Figure 3.13 for a biconvex lens, where the offset is equivalent to inserting a wedge between the curved and flat surfaces (Fig. 3.14); for small offsets, the wedge angle α is approximately equal to the edge thickness difference (ETD) divided by the lens diameter D. Table 3.4 shows typical offsets ("decentration") for various levels of lens precision, although the acceptable level must be determined not by rule of thumb but rather by including the effects of wedge in the lens designer's tolerancing as a tilt angle.

Wedge and centration specs take on different forms in a component drawing, depending on the type and use of the component. A tight centration spec, for example, is generally not required for inexpensive lenses such as those used in cell-phone cameras. In the design of prisms, however, specific (and precise) values of wedge are intentional. Similarly, for flat elements used

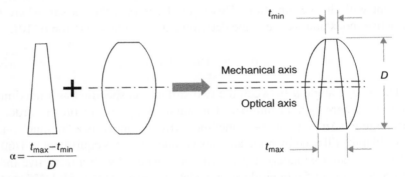

FIGURE 3.14 Element tilt deviates rays in the same manner as a prism with wedge angle $\alpha = (t_{max} - t_{min})/D = \text{ETD}/D$.

TABLE 3.4 Standard, Precision, and High-precision Tolerances Associated with Geometrical Fabrication Properties for Standard-size Optics[a] [2, 13]

Property	Standard	Precision	High precision
Center thickness	±200 μm (±0.008″)	±100 μm (±0.004″)	±50 μm (±0.002″)
Lens tilt/wedge	5 arcminute	1 arcminute	0.15 arcminute
Centration	200 μm	50 μm	10 μm
Clear aperture	95% of OD	95% of OD	95% of OD

For reference, 25.4 μm is approximately equal to a unit of length known as a "mil," or 0.001 inches; OD denotes "outer diameter."
[a] Adapted from Keith J. Kasunic, *Optical Systems Engineering*, McGraw-Hill (2011).

in laser systems, a small amount of tilt is often introduced in order to prevent Fresnel reflections from the two surfaces from interfering.

In addition to offset-induced surface tilt, the optical axis will also be tilted with respect to the mechanical axis. These tilts are difficult to quantify on a heuristic basis but when combined with decentration are critical specs in light of Kingslake's comment that "[surface] tilt does more damage to an image than any other manufacturing error" [14]. In addition, the assembly of lens elements into mechanical housings with tolerances on their diameters results in assembly tilt, a topic discussed in Chapter 4.

3.8 CLEAR APERTURE

Not all of the surface and element tolerances reviewed so far need to be satisfied over the entire size of the optic. We have seen, for example, that power and irregularity specs are given over a CA. Figure 3.15 shows that the CA is smaller than the mechanical diameter; for standard-size optics—that is, of about 5–100 mm in diameter—a typical value for the CA is 90% of the element

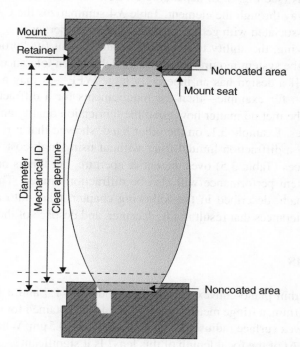

FIGURE 3.15 The clear aperture of an optical element, over which fabrication and coating quality are specified, is typically about 90% of its outer diameter. Courtesy of Edmund Optics, Inc.

TABLE 3.5 Typical Tolerances Available from a Vendor for Commercial, Precision, and High-precision Fabrication Quality[a]

Property	Commercial	Precision	High-precision
Glass index, n_d	±0.001	±0.0005	Melt data
Outer diameter, OD (mm)	+0.00/−0.10	+0.000/−0.025	+0.000/−0.015
Center thickness (mm)	±0.150	±0.050	±0.025
Clear aperture	80% of OD	90% of OD	90% of OD
Surface radius	±0.2% or 5 fringes	±0.1% or 3 fringes	±0.05% or 1 fringe
Irregularity	2 fringes	0.5 fringes	0.2 fringes
Irregularity (μm)	±10	±1	±0.5
Lens wedge (arcminute)	±5	±1	±0.5
Edge bevels (mm)	<1.0	<0.5	<0.5
Scratch-dig (MIL-13830B)	80–50	60–40	20–10
Surface finish (Å RMS)	50	20	10

[a]Data from Optimax Systems, Inc.

OD. One area for potential cost savings is that the CA can be different on each side of a lens (see Fig. 3.3), depending on how the chief and marginal rays work their way through the element. Table 3.4 summarizes the CA and other tolerances associated with geometrical fabrication properties.

Summarizing, the ability to manufacture an element (its "producibility") depends on the system requirements and the fabrication tolerances needed to meet them. If a design has an inherently large WFE—as may a fast, wide-angle singlet, for example—then the requirements for a diffraction-limited lens cannot be met no matter how good the fabrication quality and how tight the tolerances. Example 3.1, on the other hand, showed that it is possible to manufacture a diffraction-limited filter without using high-cost, high-precision tolerances (Table 3.5) over the entire aperture, but this is no guarantee that the system performance will also be diffraction-limited. The next step along that path, described in the following chapter, takes into account the assembly tolerances that result in tilt, decenter, and despace of the elements.

PROBLEMS

3.1 For a thin plano-convex lens with an index $n = 1.5$ and a focal length $f = 100\,\text{mm}$, a fringe measurement of $N = 4$ was obtained for the variation in convex surface radius ΔR at a wavelength $\lambda = 0.5\,\mu\text{m}$. What is the variation Δf of the focal length of this lens? Is it significant?

3.2 How much SFE does 3 fringes of surface irregularity correspond to, when measured with a test plate at $\lambda = 0.633\,\mu\text{m}$ and at $\lambda = 3.39\,\mu\text{m}$? If the lens being measured has an index $n=2$ at both wavelengths, what is the corresponding WFE?

3.3 What is the fabrication WFE in units of waves (PV) for the planar surface of a plano-convex lens ($n=1.5$), given the following interferogram obtained with a test plate? Note that the defect distorts the straight-line (tilt) fringes by one-half of a fringe.

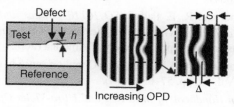

Credit: Eric P. Goodwin and James C. Wyant, *Field Guide to Interferometric Optical Testing*, SPIE Press (2006).

3.4 Is there another reason besides correlated surface errors—such as those shown in Figure 3.1—why a window would look worse reflecting a scene that it does transmitting it?

3.5 Will off-axis wavefronts be distorted for a window with correlated surface errors? Why or why not?

3.6 In Example 3.1, the measurement wavelength and design wavelength are the same ($\lambda = 633$ nm). Is the WFE for the system better or worse if the design (usage) wavelength is 1550 nm? By how much?

3.7 In Example 3.1, what are the effects on WFE of a thumbprint or localized fabrication error on the filter near the image plane? Assume the error is directly in the center of the filter. Hint: Does your answer depend on the field-of-view of the system compared with the size of the error?

3.8 Why is window wedge not a problem for the image *quality* you see looking through the windows at work or at home? Does wedge have an effect on image *location*?

REFERENCES

1. R. Williamson, *Field Guide to Optical Fabrication*, Bellingham: SPIE Press (2011).
2. D. Anderson and J. Burge, "Optical fabrication," in D. Malacara and B. J. Thompson (Eds.), *Handbook of Optical Engineering*, New York: Marcel Dekker (2001).
3. E. P. Goodwin and J. C. Wyant, *Field Guide to Interferometric Optical Testing*, Bellingham: SPIE Press (2006).

4. D. Malacara, *Optical Shop Testing* (3rd Edition) Hoboken: John Wiley & Sons, Inc. (2007).

5. D. Malacara, "Optical testing," in M. Bass, E. W. Van Stryland, D. R. Williams, and W. L. Wolfe (Eds.), *Handbook of Optics* (2nd Edition), Vol. 2, New York: McGraw-Hill (1995), Chap. 30.

6. H. H. Karow, *Fabrication Methods for Precision Optics*, Hoboken: John Wiley & Sons, Inc. (2004).

7. R. K. Kimmel and R. E. Parks, *ISO 10110—Optics and Optical Instruments: Preparation of Drawings for Optical Elements and Systems, A User's Guide* (2nd Edition), Washington: Optical Society of America (2002).

8. N. Balasubramanian and M. Hercher, "Common sense in optical specifications," Proc. SPIE, Vol. 54, pp. 57–63 (1974).

9. *The Crystran Handbook of Infra-Red and Ultra-Violet Optical Materials*, Crystran Ltd. (www.crystran.co.uk) (2008).

10. W. J. Smith, *Modern Optical Engineering* (4th Edition), New York: McGraw-Hill (2008).

11. A. B. Shorey, et al., "Surface finishing of complex optics," Opt Photonics News, Vol. 18, No. 10, pp. 14–16 (2007).

12. R. E. Parks, "Specifications: figure and finish are not enough," *Proc. SPIE*, Vol. 7071 (2008).

13. R. Parks, "Optical fabrication," in M. Bass, E. W. Van Stryland, D. R. Williams, and W. L. Wolfe (Eds.), *Handbook of Optics* (2nd Edition), Vol. 1, New York: McGraw-Hill (1995), Chap. 40.

14. R. Kingslake, *Lens Design Fundamentals*, New York: Academic Press (1978).

4

OPTICAL ALIGNMENT

Even when lenses are designed and fabricated with diffraction-limited tolerances or are purchased as off-the-shelf components of sufficient quality, their use can still result in a system that is not diffraction limited. The image degradation can be due to misalignments of the individual elements at assembly, temperature changes, and/or mechanical motion of components due to structural loads. These effects all increase wavefront errors (WFEs), which limits the optical performance (e.g., blur size) that can be obtained.

In addition to fabrication tolerances, alignment tolerances are the most critical factors in whether or not an optical system can be built at reasonable cost. When assembling an optical system consisting of multiple lens elements, unavoidable misalignments between elements include angular or linear errors resulting from assembly, temperature changes, or structural loads. As illustrated in Figure 4.1, the angular errors are known as alignment tilt; linear errors are classified as either "in the plane of the element" offsets from the optical axis (decenter), or "along the optical axis" (despace). Defocus is a special case of despace, describing the axial alignment of an optic with respect to the detector. These misalignments are reviewed in Section 4.1; the emphasis is on the effects of misalignment on image quality. Figure 4.1 also illustrates a shift in the image location—or change in line-of-sight (LOS) alignment—due to component decenter, an effect we look at in Section 4.4.

Optomechanical Systems Engineering, First Edition. Keith J. Kasunic.
© 2015 John Wiley & Sons, Inc. Published 2015 by John Wiley & Sons, Inc.

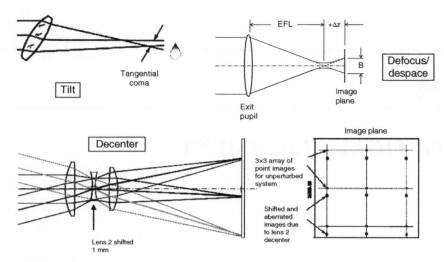

FIGURE 4.1 Misalignments may be classified as tilt, decenter, and despace. Credit for decenter graphic: J. H. Burge, Proc SPIE, Vol. 6288 (2006).

As with the fabrication of optical elements, there are tolerances on each of these misalignments and the tolerances determine the design's cost and producibility [1]. In this case, producibility refers to the ability of a machine shop to fabricate mechanical parts that maintain the required tilt, decenter, and despace (Section 4.2). The allowable tolerances are determined by the lens designer, summarized with an alignment budget, and are based on system requirements such as blur size, modulation transfer function (MTF), ensquared energy, and so on. In conjunction with the shop, the optomechanical engineer must then determine whether parts such as lens housings and spacers can be fabricated and assembled to within these tolerances; if not, then methods for obtaining the correct alignment will be required (Section 4.3). The effects of structural and thermal loads on alignment will be reviewed in Chapters 5, 6, and 7 and Chapter 8, respectively.

4.1 TYPES OF MISALIGNMENTS

Figure 4.1 shows that, in addition to an increase in aberration WFE and blur size, lens decenter will also shift the image location. The effects of component misalignment on image displacement are predictable. Unfortunately, it is difficult to predict in a simple way the effects of misalignment on aberration WFE, optical blur, and image quality—some lenses may have huge effects on WFE, while others may have little. As a result, alignment tolerances for

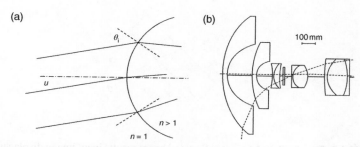

FIGURE 4.2 Low f/# (a) and wide-angle (b) systems are most susceptible to aberrations resulting from misalignments. Credit for (b): Adapted from Warren J. Smith, Modern Optical Engineering, McGraw-Hill (2008).

diffraction-limited systems are typically determined using design software such as ZEMAX, Code V, Oslo, and so on. However, some general observations can still be made. Because of the dependence of aberrations on incident angle via Snell's law:

- Fast (low f/#) systems are more susceptible to aberrations resulting from misalignments
- Wide-angle systems are also more susceptible

For fast systems with large apertures, the surface of a spherical optic forces the incident angle θ_i to be larger at larger marginal-ray heights (Fig. 2.9). That is, a tilted lens has a greater change in surface normal with tilt angle u ($\Delta\theta_i/\Delta u$) for marginal rays at the edge of a lens than it does for paraxial rays near the optical axis (Fig. 4.2a). Lens tilt can thus have a larger effect on the comatic aberration of a fast (large aperture) lens than it does on a slow (small aperture) lens. Lens designers use a number of techniques for reducing alignment sensitivities, such as splitting elements to reduce the radii of each lens—the same techniques that are used for reducing aberrations. Similarly, wide-angle lenses with large incidence angles are more sensitive to changes in incident angle Δu (Fig. 4.2b). Additional details for tilt misalignments, as well as decenter and despace, are given in the following sections.

4.1.1 Tilt

The simplest of optical elements—the flat (or *plane parallel*) plate—tilted in an unfocused beam displaces the beam from the optical axis (Fig. 4.3). The displacement $d \approx t\theta_i(n-1)/n$ for small angles [2], which depends on the refractive index n of the plate, the plate thickness t, and the incident angle θ_i.

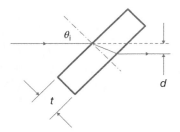

FIGURE 4.3 Tilt of a flat plate in a collimated beam results in an image displacement d.

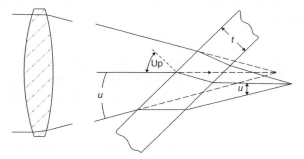

FIGURE 4.4 Tilt of a flat plate in a focused beam results in both aberrations and an image displacement. Adapted from Warren J. Smith, Modern Optical Engineering, McGraw-Hill (2008).

Physically, a larger incident angle bends the incident ray more (via Snell's Law), while a thicker plate has a longer distance for the bent ray to propagate. This effect is used, for example, by astrophotographers to correct for the offsets of stars caused by the atmosphere. That is, air turbulence creates small variations in the atmosphere's refractive index, changing the position at which a star appears in the sky. It is possible—and commercial products are available—to compensate for this offset with a flat plate tilted about two axes (known as "tip-tilt" axes), thus keeping the star on the same position on the detector array. Such adaptive-optics techniques are used to increase the resolution of star imagery.

For a flat plate tilted in a focused beam, beam displacement still occurs, but aberrations which increase the optical blur size also occur (Fig. 4.4). These aberrations depend on the tilt angle u_p and cone angle u for the focused beam. The simplest element in optics thus now has spherical aberration, coma, and astigmatism that depend on the refractive index and cone angle (for spherical aberration), as well as the tilt angle for coma and astigmatism [2].

The tilt of a lens or curved mirror may deviate the optical axis from its intended direction; the image displacement for thin lenses is extremely small [2].

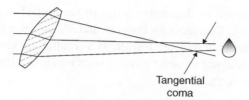

Tangential
coma

FIGURE 4.5 Tilt of a thin lens results in an increase in blur size known as coma, with almost no change in pointing angle. Adapted from Warren J. Smith, Modern Optical Engineering, McGraw-Hill (2008).

A more important effect is that the wavefront incident on a tilted lens surface results in coma or astigmatism (Fig. 4.5). Unfortunately, there is no general method to predict the effects of tilt on blur size and image quality. When the misalignment is such that the blur is larger than diffraction limited, we will see in Section 4.4 that a powerful method exists for such predictions. In most cases, however—that is, where the peak-to-valley WFE must be held to $<\lambda/4$—this method is not valid, and Monte Carlo tolerancing and diffraction-analysis tools with software packages such as Zemax and Code V must be used instead.

4.1.2 Decenter

We have seen in Chapter 3 that the fabrication error known as wedge can result from the decenter between the optical and mechanical axes of a lens. In addition to being a fabrication tolerance, decenter is also an alignment tolerance; in this case, it describes the parallel offset between the optical axes of different elements when they are assembled. As illustrated in Figure 3.14, these axes are defined by the centers of curvature of each lens. The alignment offset between axes is the result of small clearances between the lenses and the mounts that retain them. The offset can be parallel to both the x and y axes; thus decenter is a result of lateral xy-misalignment with respect to the relative placement of the optic in the xy-plane perpendicular to the nominal optical axis z.

Centration will often be the most important alignment requirement, as decenter of a powered lens or mirror results in a shift in image location. As with tilt, the asymmetry induced by decenter also results in coma in the image; on-axis coma is a typical symptom [3]. The acceptable decenter is determined by the lens designer, with the steeper ray angles associated with fast, wide-field optics again having tighter tolerances. An important exception is the decenter for flat surfaces, which has no effect on WFE and image quality because offsets do not affect wavefront incident angles on these surfaces. By contrast, curved surfaces are affected by decenter, with off-axis elements being more sensitive to decenter owing to the initial asymmetry of the ray-intercept angles. Tolerances for both on- and off-axis elements are generally tighter for decenter than they are for tilt; as a result, they must also be specified by the lens designer in order to meet such imaging requirements as blur size, MTF, and so forth.

4.1.3 Despace

While optical components can be shifted laterally in the xy-plane with respect to each other, they can also be misaligned along the optical axis (Fig. 4.6). Such variations from the nominal lens-to-lens spacing L are known as *despace*. That is, if the spacing L is not accurately set, there will be some variation in the expected EFL (system focal length f_{sys}) due to this despace ΔL; the tolerances on despace may be determined by the change in focal length that results. In addition, the changes in L may also change slightly the incidence heights of the rays on each lens, thus affecting the image aberrations. For the two-lens system shown in Figure 4.6, for example, moving the two lenses slightly closer together increases the height of the rays incident on the second lens, which—as we have seen in Figure 2.12—increases spherical aberration and may also affect coma and astigmatism for off-axis points [4].

4.1.4 Defocus

At some point, wavefronts propagating through an optical system will exit the last lens and come to focus on a single-pixel detector or FPA. If the detector is not placed at the correct distance from the lens, then images will be blurred, and all the effort that has gone into designing, fabricating, and aligning a

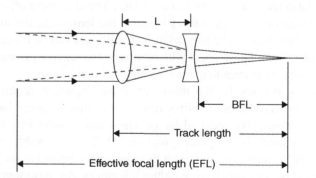

FIGURE 4.6 The spacing L between elements determines the EFL of the lens assembly. A despace ΔL changes the EFL and image-quality aberrations.

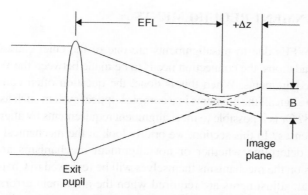

FIGURE 4.7 The depth of focus for the distance between the exit pupil and image plane for a diffraction-limited lens is based on wavefront diffraction (solid lines). The figure illustrates Equation 4.1 with the "+" sign; with the "–" sign, the image plane would be to the left of the focus by a distance Δz. Adapted from Keith J. Kasunic, Optical Systems Engineering, McGraw-Hill (2011).

diffraction-limited lens will not produce the crisp image expected. Fortunately, the detector need not be located an exact distance from the lens, and the tolerance with respect to this distance is known as depth of focus (DoF) or allowable defocus—that is, the distance over which the blur size remains acceptable (Fig. 4.7).

The diffraction-based defocus is based on the Rayleigh criterion of $\lambda/4$ PV wavefront error. The DoF Δz depends on the wavelength—which is expected for a diffraction-based criterion; it also depends on the f/# [2]

$$\Delta z = \pm 2\lambda \, (f/\#)^2 \quad [\mu m] \tag{4.1}$$

This DoF is based on the Rayleigh limit, where a quarter wavelength of peak-to-valley WFE produces a focus shift of Δz. This means that design and fabrication efforts will be wasted if the lens isn't focused properly or if the FPA itself isn't sufficiently flat. The distance from the lens to the FPA can sometimes be controlled with a focus mechanism; the value obtained from Equation 4.1 indicates how difficult it would be for such a mechanism to work. Typical numbers based on this criterion for an $f/2$ lens using visible-band optics are $\Delta z = \pm 2 \times 0.5 \, \mu m \times 4 = \pm 4 \, \mu m$, a value that increases linearly with wavelength to $\pm 80 \, \mu m$ for LWIR optics at a wavelength of $10 \, \mu m$.

4.2   ALIGNMENT REQUIREMENTS

Given that WFEs due to misalignments are one of the chief causes of image-quality degradations, the connection needs to be made between the WFE and the alignment requirements. When this is done, the question often comes up: is it necessary to align all the elements in a complex optical system such as that shown in Figure 4.8? Or is it possible to meet alignment requirements by aligning a small number of lenses? In this section, we briefly look at the mechanical tolerancing which will determine whether or not alignment mechanisms are required; more details on the mechanisms themselves will be reviewed in Chapter 9.

In general, adjustments are required when the alignment errors that result from assembling without adjustments exceed the allowable tolerances. Tilt, decenter, despace, and defocus will all be affected by the mechanical toler-ances of fabricated parts. Looking first at tilt, a lens spacer can be machined with a moderate-cost wedge (parallelism) of $12.5\,\mu m$ (0.0005 in.) over a 50 mm diameter (Fig. 4.9). This results in a tilt angle θ_t between the lenses of

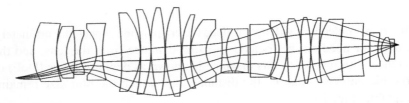

FIGURE 4.8 Complex lens assemblies contain many elements, most of which cannot be adjusted. Adapted from Robert Fischer et al., Optical System Design, McGraw-Hill (2008).

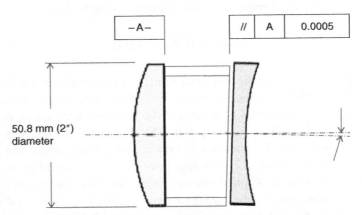

FIGURE 4.9 The angular tolerancing of the spacer separating two lenses determines the assembly tilt. The two angled parallel lines in the box on the upper right indicate that the surface is to be parallel to surface A (datum "−A−") within 0.0005 in. ($12.7\,\mu m$).

12.5 μm/0.05 m = 250 μrad. This is somewhat more than the lens wedge requirement of 200 μrad, for example, indicating that spacer tilt may need to be reduced for this example. Other situations where the assembly tilt significantly exceeds the lens wedge may require an alignment mechanism.

Alignment decenter determines the allowable mechanical tolerances when fabricating lens and housing diameters. When the clearance between the lens diameter and the inner diameter of the housing is too large, for example, the resulting decenter will degrade image quality; a standard tolerance (and thus decenter spec) for this clearance is on the order of 0.05 mm (~0.002 in.). For example, lens diameters can be machined with a moderate-cost tolerance of 25–50 μm (0.001–0.002 in.) over a 50 mm diameter (Fig. 4.10). As the lens diameter is determined by a grinding process, it cannot generally be held to as tight a tolerance as a machined housing, shown in Figure 4.10 as a lens cell having a tolerance of +12.5 μm/−0.0 μm (+0.0005 in./−0.0 in.). The tolerance stackup determining decenter is thus 25 μm (0.001 in.) nominal, and 62.5 μm (0.0025 in.) worst case. This may or may not exceed the allowable decenter established by the lens designer; if it does, tighter—though more expensive—tolerances are available for the machining of the lens cell.

Finally, the mechanical tolerances determining despace and defocus are illustrated in Figure 4.11. Lens spacers can be machined with a moderate-cost thickness tolerance of ±12.7 μm (±0.0005 in.). Shims as thin as 5 μm (0.0002 in.)

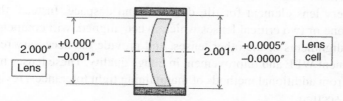

FIGURE 4.10 The tolerance on the diameter of both the lens and its housing determines the decenter.

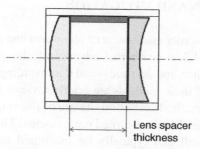

FIGURE 4.11 The tolerance on the thickness of the lens spacer determines the despace tolerance.

TABLE 4.1 The Tolerancing of Various Machined Features Determines the Tilt, Decenter, and Despace[a]

Parameter	Commercial	Precision	High precision
Cost	Low	Moderate	Expensive
Spacing—manual (µm)	200	25	6
Spacing—CNC (µm)	50	12	2.5
Concentricity—multiple chucks (µm)	200	100	25
Concentricity—single chuck (µm)	200	25	5
Parallelism (µm)	25	12	5
Lens-cell diameter (µm)	±100	±25	±6

[a]Data courtesy of David Anderson and Jim Burge, "Optical Fabrication" in The Handbook of Optical Engineering, Chapter 28.

are also available, and can be used for more precise control of despace, but are difficult and expensive to use as they are labor intensive.

If the moderate-cost tolerances on tilt, decenter, and despace exceed the requirements established by the lens designer for each lens, then tighter tolerances may be necessary (Table 4.1), or it may be necessary to use an alignment mechanism. It is not always clear which of these options is the least expensive, nor which lenses may need to be aligned, and to what extent. In general, if an alignment mechanism is required, it *never* makes sense to align every lens element for tilt, decenter, and despace. Instead, there are usually one or two critical lenses which, when aligned, will compensate for the misalignments of the other lenses, and provide the necessary reduction in alignment WFE and improvement in image quality. These lenses may also benefit from additional methods of maintaining tight tolerances, reviewed in the next section.

4.3 CORRECTION AND MITIGATION

In addition to the first-order assessment of alignment and machining tolerances reviewed in Section 4.2, a number of advanced techniques are available for minimizing tilt, decenter, and despace—and thus avoiding the use of alignment mechanisms. Some of these methods are briefly reviewed in this section.

The first of these methods relies on the use of the smaller mechanical tolerances available from diamond-turned components. That is, the mechanical diameter of a lens cell can typically be machined to a tighter tolerance than the outer diameter (OD) to which a lens can be ground—see Table 4.1. The exception is when hardened, diamond-tipped tools are used to machine

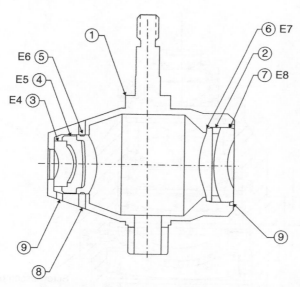

FIGURE 4.12 Surface 3 of lens element 4 (E4) was diamond-turned with tight tolerances to allow low-decenter assembly with lens element 5 (E5), which was similarly
machined. Robert Guyer et al., Proc. SPIE, Vol 3430, pp. 109–113 (1998).

the lens OD; in this case, the OD can be machined to tolerances as tight as
that of the metal housing, and the accumulation ("stackup") of worst-case
tolerances between the lens OD and lens-cell diameter may then be sufficiently small to meet tight decenter requirements. An example is shown in
Figure 4.12, where IR lenses were diamond-turned with positioning features
(edges and flats) located to within a few microns of the optical axis. Not all
materials can be diamond machined—aluminum mirrors and germanium,
zinc selenide, and zinc sulfide lenses are materials that can be, but glasses
cannot—and in the cases where they can be, it may be possible to assemble
lens elements without the use of lens housings. Additional design factors
will also come into play in making such a decision—see the discussion in
Chapter 8, for example, on thermal expansion coefficients.

When diamond turning is not possible, the second method is based on
using the mechanical tolerances given in Table 4.1, and allowing a critical
lens to be adjusted. When the clearance between the lens diameter and the
inner diameter of the housing is too large, for example, the resulting
decenter will degrade image quality; a standard tolerance (and thus decenter spec) for conventional machining of this clearance is on the order of
50 μm (~0.002 in.). Another tolerance that creates decenter is the lack of
concentricity between different lens assemblies, which results when the

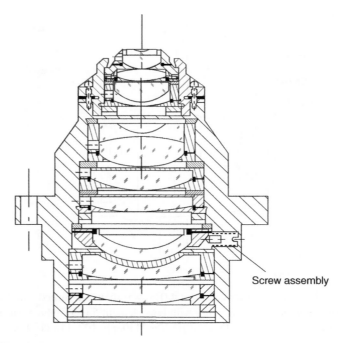

FIGURE 4.13 The mounting of lens elements centered in individual lens cells allows a reduction in decenter stackup in complex assemblies. Credit: Daniel Vukobratovich, Optomechanical System Design, in The Infrared and Electro-Optical Systems Handbook, Vol 4, Chapter 3, SPIE Press (1993).

subassemblies are attached together with flanges or threads; this tolerance can be alleviated by mounting the assemblies (or "lens cells") in the same housing (Fig. 4.13). The tolerance stack-up—of each lens in its lens cell and the lens cells in the housing—contributes to decenter, though this is minimized by the use of a post-adjustment bond that allows decenter adjustments during assembly, and by the ability to hold tighter tolerances on the diameter of the metal lens cell than on the lens itself. Final decenter alignment and image quality is set by adjusting the highly curved lens with the screw assembly shown in Figure 4.13.

More generally, four approaches are available for controlling the centration of lenses and subcells, which is typically the misalignment to which lenses are most sensitive [3]. In order of precision and cost (low to high), they are (i) shims, (ii) pin-and-pot, (iii) optical centering, and (iv) interferometers.

Due to their lack of stiffness, thin shims are very difficult to use, and a technique known as "pin-and-pot" is more common for centration alignments.

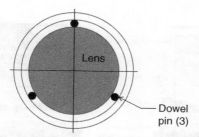

FIGURE 4.14 Precisely machined dowel pins control lens decenter in a method known as "pin-and-pot." Pin diameters (and thus a bond thickness) of <1 mm are available.

This approach uses ground dowel pins at three equally spaced places around the circumference of the lens (Fig. 4.14). Dowel pins can easily be machined to diameter as small as 0.039 in. (1 mm), a dimensional tolerance of 0.0002 in. (5 μm), and are much easier to use than shims. After the pins are located, the lens is potted in place with a room-temperature vulcanizing (RTV) adhesive. To prevent damage to the optic, the pins are removed after the RTV has set. Tolerance stackup is good to nearly 0.001 in. (25 μm) of decenter.

Even tighter decenter can be obtained using a technique known as optical centering. With this approach, reflections from individual lens surfaces are used to determine the decenter of their center of curvature. This requires rotation of the lens about a precisely known reference axis. Equipment is available commercially for such alignments from companies such as TriOptics GmbH,[1] allowing decenter as small as 0.0002–0.0004 in. (5–10 μm) after epoxy or RTV cure.

For even smaller decenter, interferometers can be used as an expensive, time-consuming method for low-volume monitoring of decenter during alignment. For high-volume production optics, increasing the clearances between mating parts allows more degrees of freedom for aligning the optics. For example, the conventional threaded housing shown on the left of Figure 4.15 can be replaced with the design on the right, allowing tilt and decenter adjustments by directly monitoring the image quality with the image sensor before final adhesive bonding.[2]

Finally, a fourth approach to alignment management is to use optical components that are relatively insensitive to misalignments. These include elements such as retroreflectors, porro prisms, and pentaprisms—reflective components typically used in laser systems. An example is shown in Figure 4.16, where two porro prisms are used to define a laser cavity, and a retroreflector is used to fold the cavity into a shorter length. The design was able to tolerate angular misalignments up to 100 μrad—a very large number for a complex laser system [5].

[1] See, for example, the OptiCentric brochure on the TriOptics GmbH website http://www.trioptics.com.

[2] See, for example, the Automation Engineering, Inc. web site http://www.aeiboston.com.

(a)   (b)

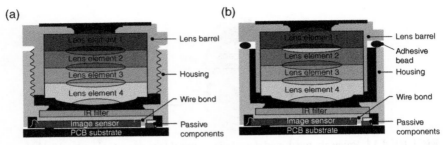

FIGURE 4.15 Tilt and decenter alignments of a lens assembly can be monitored directly with the image sensor before adhesive bonding. Adapted from: Andre By, Vision Systems Design, April 2011, pp. 12–15.

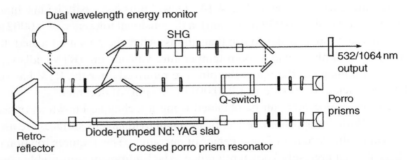

FIGURE 4.16 Both porro prisms and retroreflectors are used to design lasers which are relatively insensitive to misalignments. Credit: Floyd Hovis, Proc SPIE, Vol. 6100 (2006).

4.4 POINTING AND BORESIGHTING

The alignment methods looked at in the previous sections emphasized their effects on image quality. That is, the tilt, decenter, despace, and defocus of optical elements were all to be reduced to the level where diffraction-limited image quality could be obtained. As shown in Figure 4.1; however, misalignments such as decenter can also result in an image shift or displacement. The purpose of this section is to review the degree of displacement expected for a given misalignment, as well as the hardware available to correct for these displacements.

Image displacements can be broadly categorized as one of two types. The first results in the situation where the mechanical pointing of the optical system is not the same direction as where the object appears to be located. A simple example is illustrated in Figure 4.17, where the wedge angle of a thin prism refracts the rays from a distant star as they propagate through the prism. The angular deviation $\delta \approx (n-1)\alpha$, where α is the prism angle due to fabrication tilt or wedge in a lens or window; the resulting error is known as LOS or pointing misalignment.

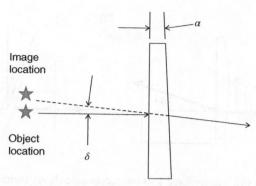

FIGURE 4.17 A thin prism with a tilt angle α—a flat plate fabricated with wedge, for example—refracts light such that an object appears to be at a different position than its actual location.

The second type of image deviation occurs when there are multiple instruments or optical systems that need to be aligned with respect to each other, to insure that they are all pointing at the same object within some small degree of error. This relative pointing angle is known as *boresight error*, although LOS misalignment within each instrument is sometimes known as boresight error as well.

As with image-quality changes, the LOS misalignment can be due to misalignments of the individual elements at assembly, temperature changes, and/or mechanical motion due to structural loads. The analysis of pointing misalignments starts by recognizing that despace or defocus does not cause an image shift in the plane of the detector; as we will see, image shifts occur due to decenter, which redirects rays across the image plane. In Figure 4.18, for example, we see the effects of tilt and decenter of a perfect thin lens with no wedge. The figure shows that the decenter of a thin lens introduces a nonzero Snell's law angle for the on-axis ray which now intercepts the lens off the optical centerline. As a result, the ray is bent at an angle $\Delta\theta$, giving an image displacement in the detector plane of $\Delta y = f\Delta\theta$. Similarly, Figure 4.19 shows the effects of a decentered or tilted mirror.

Equation 4.2 shows that the LOS deviations in the image plane (ε_i) depend on the angular deviation $\Delta\theta$ that *results* from the misalignment [6, 7].[3]

$$\varepsilon_i \cdot NA_{sys} = \frac{1}{2}\Delta\theta \cdot D_{beam} \quad [mm-rad] \quad (4.2)$$

---

[3] For those familiar with the concept, the units of Equation 4.2 hint at the idea of the Lagrange Invariant. This concept—and the related idea of etendue—was used to derive Equation 4.2; see Ref. [7] for more details.

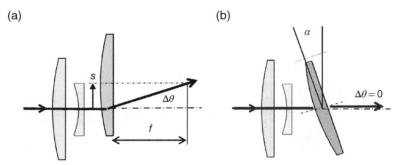

FIGURE 4.18 In (a), a decentered lens intercepts on-axis rays with non-zero Snell's law angles, bending the rays with an angular deviation $\Delta\theta = s/f$. In (b), rays continue to go through the center of a tilted thin lens with little deviation. Credit: J. H. Burge, Proc. SPIE, Vol. 6288 (2006).

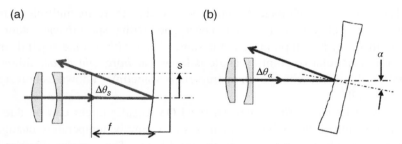

FIGURE 4.19 In (a), a decentered mirror intercepts on-axis rays with nonzero Snell's law angles, bending the rays with an angular deviation $\Delta\theta_s = s/f$. In (b), rays reflect off a mirror tilted at an angle α with an angular deviation $\Delta\theta_\alpha = 2\alpha$. Credit: J. H. Burge, Proc. SPIE, Vol. 6288.

Looking at Figures 4.18 and 4.19, there are three possible causes: decenter of a thin lens or mirror (for which the angular deviation is $\Delta\theta$ in Fig. 4.18 or $\Delta\theta_s$ in Fig. 4.19), or tilt of a mirror (for which the angular deviation is $\Delta\theta_\alpha$ in Fig. 4.19).

Equation 4.2 also shows that the image shift ε_i depends on the numerical aperture (NA) of the system [where $NA_{sys} = 1/2(f/\#)_{sys}$—see Fig. 4.20], as well as the diameter of the beam incident on the lens (D_{beam}). As a simple example of the use of Equation 4.2, a mirror decentered by an amount s will result in a ray deviation $\Delta\theta = s/f$; for a flat mirror with no power, $f = \infty$ and $\Delta\theta = 0$, from which Equation 4.2 gives $\varepsilon_i = 0$ as expected. Both tilt and decenter can of course also be combined with Equation 4.2, as the ray deviations are independent and superimpose for small angles. Example 4.1 illustrates the use of the equation for a two-lens system.

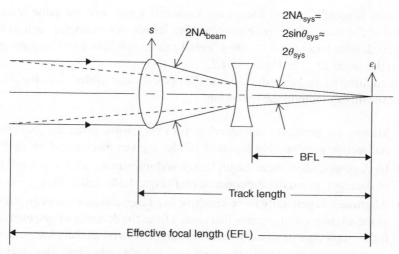

FIGURE 4.20 A telephoto lens whose first element is decentered by an amount s will result in an image shift ε_i on the focal plane.

Example 4.1   As shown in Figure 4.20, a simple telephoto lens consists of a positive and negative element, giving an effective focal length (EFL) that is longer than the track length, and allowing close-up images of distant objects. Also shown are the numerical aperture angle NA_{lens} of the decentered lens, as well as a numerical aperture angle NA_{sys} of the two-lens system. How much image displacement ε_i will result when the positive lens is decentered by an amount s?

From Equation 4.2, an offset s results in an angular change $\Delta\theta$ such that $\varepsilon_i \cdot NA_{sys} = 0.5 \ \Delta\theta \cdot D_{lens}$, where $D_{beam} = D_{lens}$ for the first lens. Substituting for $\Delta\theta = s/f_{lens}$ gives $\varepsilon_i \cdot NA_{sys} = 0.5s \cdot D_{lens}/f_{lens}$. For a telephoto lens with $L = 50\,mm$, $f_1 = +100\,mm$, and $f_2 = -150\,mm$, the $EFL = (1/f_1 + 1/f_2 - L/f_1 f_2)^{-1} = 150\,mm$. For an aperture diameter $D_{lens} = 20\,mm$ ($y = 10\,mm$), this gives $NA_{lens} = y/f_1 = 10\,mm/100\,mm = 0.1\,rad$ and $NA_{sys} = y/EFL = 10\,mm/150\,mm = 0.06667\,rad$.   With $2NA = 1/(f/\#) = D/f$, we find $\varepsilon_i = s \cdot NA_{lens}/NA_{sys} = s \cdot (0.1/0.06667) = +1.5s$. Physically, the large EFL (small NA_{sys}) of the system magnifies the effects of the decenter s, compared with a simple thin lens by itself.

The sensitivities of a system to decenter or tilt depend on the type of optic, and where it is located in the system. For example, the on-axis beam diameter D_{beam} incident on the second lens in Example 4.1 is clearly smaller than the beam diameter incident on the first lens; Equation 4.2 then shows that the second lens is thus inherently less sensitive to decenter, and has looser tolerances associated with its assembly for LOS pointing. In the extreme case—as with field lenses and field flatteners located near an image—the beam diameter is small, and the decenter tolerance is loose. For the same reason, lenses at the front of the system—objective lenses, for example, such as the positive lens in Example 4.1—are generally more sensitive to misalignments, given the larger D_{beam} in Equation 4.2.

In addition to the beam diameter at the optics, some useful rules for alignment sensitivity include the following:

- Mirrors are generally "touchier" to tilt than lenses, given the 2× optical redirection illustrated in Figure 4.19b for a given mechanical tilt angle.
- High power, short focal length lenses and mirrors are also less forgiving of decenter, given the dependence on f (Figs. 4.18a and 4.19a).
- Alignment is generally more sensitive to a large distance between lenses, as the angular misalignment that results from the decenter of one element, for example, is magnified over a long length. As a result, keeping elements close together in a small, compact package not only saves size, weight, and cost, but it also improves manufacturability and producibility through looser alignment tolerances.
- Example 4.1 illustrated the dependence of ε_i on component f/#, so the highly curved surfaces and high-index materials leading to small f/# have tighter alignment requirements. Alternatively, such lenses can be used for final alignment, as illustrated by the "screw assembly" in Figure 4.13.

These rules are summarized in Table 4.2 for a variety of optical components.

As reviewed in previous sections, misalignments can first be controlled with the machining tolerances of the mounting hardware for the optics.

TABLE 4.2 Alignment Requirements and Tolerances for Various Optical Components. The Tolerances Shown Generally Apply to Image Quality and LOS Pointing; However, LOS Pointing is Relatively Insensitive to Lens and Window Tilt

Type of component	Alignment requirements	Alignment tolerance
Window	Tilt	Very loose
Field flattener, field lens	Tilt, decenter, despace	Loose
Objective lens	Tilt, decenter, despace	Moderate (tilt, despace) Tight (decenter)
Relay lens	Tilt, decenter, despace	Moderate (tilt, despace) Tight (decenter)
Flat mirror	Tilt, despace	Loose (despace) Tight (tilt)
Curved mirror	Tilt, decenter, despace	Very tight

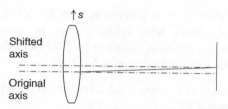

FIGURE 4.21 If a camera is moved by platform or hand motion (not shown), lens decenter *s* can be used to maintain LOS stabilization by keeping the image centered on the FPA.

When these are insufficient—as they might be for curved mirrors being used in a vibration environment which dynamically alters the alignment—then other approaches are required to maintain LOS pointing and image quality requirements. These include the tilting of a mirror located in a strategic position [4]. On occasion, decenter is used to counterbalance motions of the image on the detector. Figure 4.21, for example, shows the situation where a camera has been unintentionally moved by, say, satellite vibrations or an unsteady hand; the lens (or sometimes the detector) is then shifted to compensate, to keep the image on the center of the detector. In other situations—typically laser systems with a small beam diameter, such as that shown in Figure 4.16—a thin prism such as that shown in Figure 4.17 is rotated about its axis to control the LOS pointing. Such stabilization methods all rely on the structural stiffness of the mechanical components being sufficient to minimize tilt, decenter, and despace of the optics—a topic covered in detail in Chapter 5.

PROBLEMS

4.1 We know from Chapter 3 that a flat plate will always be fabricated with some degree of wedge. What are the effects of wedge on the tip-tilt adaptive-optics compensation of atmospheric turbulence?

4.2 Would you expect the tilting misalignment of a lens with a short focal length to have more or less WFE than a lens with a long focal length? Hint: How do the incidence angles on the lens surfaces change with focal length?

4.3 Would you expect the tilting misalignment of a lens with a short focal length to have more or less effect on the WFE of a lens following this lens, in comparison with a lens with a long focal length?

4.4 If spherical aberration is proportional to D^3 (D = the beam diameter incident on the lens), what is the change in spherical aberration if despace decreases the beam diameter by 1%? Does the despace increase or decrease the spherical aberration?

4.5 What is the effect of a despace $\Delta L = 100\,\mu m$ on the EFL of the telephoto lens in Example 4.1?

4.6 What is the image shift ε_i for a simple thin lens decentered by an amount s? Is your result consistent with physical reasoning?

4.7 How much will the image from a telephoto lens shift when the second (negative) lens is decentered by an amount s? If it is different from that found in Example 4.1 for decenter of the first (positive) lens, explain why. Hint: What is D_{beam} for the second lens?

4.8 How would you use Equation 4.2 to analyze the situation where the tilt of a thin lens is not about the centerline (as shown in Fig. 4.18), but about the edge? Hint: any rigid-body motion can be thought of as a sum of a tilt and a displacement.

REFERENCES

1. R. L. Fisher, "Incorporation of assembly and alignment methods at the design stage of complex optical systems," Proc. SPIE, Vol. 330, pp. 2–4 (1982).

2. W. J. Smith, *Modern Optical Engineering*, London: McGraw-Hill (2008).

3. R. Kingslake, *Lens Design Fundamentals*, New York: Academic Press (1978).

4. K. J. Kasunic, *Optical Systems Engineering*, New York: McGraw-Hill (2011).

5. F. Hovis, "Qualification of the laser transmitter for the CALIPSO aerosol lidar mission," Proc. SPIE, Vol. 6100, p. 61001X (2006).

6. K. Schwertz and J. H. Burge, *Field Guide to Optomechanical Design and Analysis*, Bellingham: SPIE Press (2012).

7. J. H. Burge, "An easy way to relate optical element motion to system pointing stability," Proc. SPIE, Vol. 6288, p. 62880I (2006).

5

STRUCTURAL DESIGN—
MECHANICAL ELEMENTS

Structural design is often thought of as belonging to the realm of civil engineers, builders of bridges and skyscrapers [1]. The transfer of these skills to optomechanical design requires a change in thinking—from designs that don't collapse, to designs that don't distort. While breakage due to excessive mechanical forces is always a concern, controlling the distortion and misalignment of structures such as those shown in Figure 5.1 is the optomechanical engineer's primary structural design goal. In the design of such structures, using lightweight cardboard and bubble gum as structural elements would result in misalignments; using steel and titanium fasteners, however, likely will not. But steel and titanium are relatively heavy, so how can optical alignments be maintained without exceeding weight requirements?

The design of lightweight structures for optical systems addresses two critical factors: (i) the distortion of optical elements due to applied forces, which increases the surface figure error; and (ii) structural deformation that changes the tilt, decenter, despace, and defocus alignments between elements. Even the most elegant optical design, for example, will not work if an undistorted mirror is mounted at the end of a long, wobbly structure. Thus, the structural design of optical systems—which are seldom exposed to forces sufficient to challenge strength—is concerned primarily with stiffness and deflection.

Optomechanical Systems Engineering, First Edition. Keith J. Kasunic.
© 2015 John Wiley & Sons, Inc. Published 2015 by John Wiley & Sons, Inc.

FIGURE 5.1 Controlling the distortion and misalignments of an optical system using lightweight elements is the primary goal of optomechanical structural design. Credits: Planewave Instruments (left); Alluna Optics (right).

This chapter reviews the basic concepts needed to prevent excess structural deflection of the mechanical components holding the optics in alignment. Alignment tolerances can be maintained by proper selection of structural geometry, design, and materials. The forces in this chapter are all assumed to be "static"—that is, not changing with time. In Chapter 7, we look in detail at how a structure may experience greater deflection when an applied force is changing than when it is not. Such variable (or "dynamic") forces are an inevitable part of all optical systems and they, too, affect alignment and component surface figure. The distortion of the optical components themselves is addressed in Chapters 6 and 9, and mainly concerns element size, shape, material, and mounting techniques.

5.1 STRESS, STRAIN, AND STIFFNESS

All materials have an inherent springiness or stiffness such that they stretch… they bend…they twist…and eventually they break. This stiffness is due to interatomic forces. Different materials—steel, aluminum, glass, plastic, and so on—thus bend, twist, or stretch differently. This and the following sections will first quantify this stiffness, and then use the concept to introduce structural components useful in optomechanics.

As shown in Figure 5.2, a force P stretches a bar, with a larger force stretching the bar more. In addition, a longer length L has more atoms in the "chain," with the total stretch ΔL proportional to the number of atoms. As a result, the change in length depends on both the applied load and the initial length

(a)

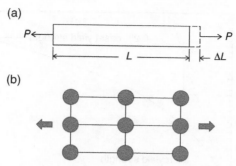

(b)

FIGURE 5.2 In (a), the stretch ΔL of a beam depends on the applied force P and the initial length L of the beam. Physically, stretching more atomic bonds—as in (b)—results in a bigger ΔL.

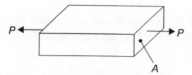

FIGURE 5.3 The stress in a beam depends on the applied force P and the cross-sectional area A.

$(\Delta L \sim PL)$. To compare different lengths of material, a "normalized stretch" is defined (symbol ε), given by $\varepsilon = \Delta L/L$ and called *strain*.

While a larger force stretches the bar more, a smaller cross-sectional area A—with fewer atoms to resist the force—allows the same (Fig. 5.3). As a result, the change in length depends on both the applied load and the cross-sectional area $(\Delta L \sim P/A)$. To fairly compare different areas, a normalized force is defined (symbol σ), given by $\sigma = P/A$ and called *stress*. With units of pounds per square inch (psi) and N/m^2, stress can be thought of as an internal pressure due to externally applied forces.

Putting these concepts together, the change in length depends on the force, length, and area $(\Delta L \sim PL/A)$, or $\varepsilon \sim P/A \ (=\sigma)$. But nothing said so far about stress or strain takes into account the stiffness of the different materials that the bar might be made of. Intuitively, we expect a stiffer bar to strain less than a flimsy one. Plotting stress versus strain, we find that they are proportional over a limited range (Fig. 5.4). As a result, the strain increases with the stress and decreases with the stiffness, such that [2]

$$\varepsilon = \frac{\sigma}{E} = \frac{P}{AE} \rightarrow \Delta L = \varepsilon L = \frac{PL}{AE} \qquad (5.1)$$

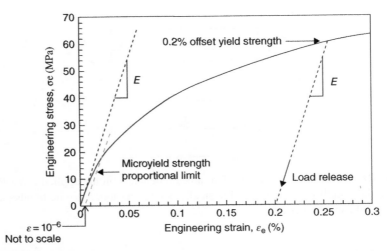

FIGURE 5.4 As a beam is strained, the stress increases linearly over a limited range. Credit: Prof. David Roylance, MIT OpenCourseWare 3.11 Course Notes.

The constant of proportionality between stress σ and strain ε is known as the elastic modulus E. It is also known as Young's modulus, named after the English researcher Thomas Young.

The elastic modulus is thus a measure of material stiffness in resisting tension, pulling, and stretching. A larger number indicates a material has little flexibility when resisting forces; it is a result of larger interatomic spring forces. Physically, the strain is smaller when there are greater bonding forces between atoms (larger E). Typical numbers are $E \approx 10 \times 10^6 \, \text{psi} \, (\text{lb/in}^2) = 69 \times 10^9 \, \text{Pa}$ (N/m^2) for aluminum and $E \approx 30 \times 10^6 \, \text{psi} = 207 \times 10^9 \, \text{Pa}$ for steel, using the conversion factor of 6895 Pa/psi.

The linear range over which stress and strain are proportional is known as the *elastic range*, over which $\sigma = \varepsilon E$ (a microscopic version of *Hooke's Law* for linear springs, where $F = kx$). Permanent deformation occurs once the strain exceeds this elastic ("proportional") limit, where $\sigma < \varepsilon E$ (Fig. 5.4). The conventional criterion for where the elastic limit is exceeded—typically used by designers of automobiles, tractors, and the like—is 0.2% strain ($\varepsilon = 0.002$, or 2 parts/1000). That is, after the elastic limit is reached and the straining force is released, there will be 0.2% of permanent strain remaining in the material. The stress which creates this strain is known as the *yield stress* or *yield strength* ($\sigma_Y \sim 60 \, \text{MPa}$ in Fig. 5.4). For optomechanical structures where the displacements and strains are smaller by orders of magnitude, a 0.2% strain is excessive and results in unacceptably large permanent misalignments between optical elements. As a result, a microstrain criterion is used instead:

$\varepsilon = 10^{-6}$ [or 1 microstrain, which equals 10^{-6} m (1 μm) of elongation ΔL per meter of initial length L] is the maximum permissible strain after load release for optical systems [3, 4]. The stress that creates the microstrain is the *microyield strength* ($\sigma_{MYS} \sim 18$ MPa in Fig. 5.4), also known as the *precision elastic limit*.

Example 5.1 An aerial-survey camera is collecting images from an altitude of 3 km. An *f*/2 lens with a 1-m focal length is required for the resolution needed to exceed that of Google Earth's camera. The problem is:

1. How much will the lens structure stretch when pointed down?
2. Does this exceed the depth of focus of the lens?

The lens weighs 44.5 N (10 lb). Assume it is mounted in an aluminum tube with a wall thickness of 1 mm (0.04 in.)—see Figure 5.5.

How much will the lens structure stretch (defocus) when pointed down? Using $\Delta L = PL/AE$, we know P, L, and E, but need the area A:

$$A = \frac{\pi}{4}(D_o^2 - D_i^2)$$
$$= \frac{\pi}{4}\left[(0.502 \text{ m})^2 - (0.500 \text{ m})^2\right] = 0.001574 \text{ m}^2$$

Note that the diameter was determined from the fact that, for an f/2 lens with $f = 1000$ mm, the inner diameter $D_i = f/(f/\#) = 1000$ mm/2 = 500 mm. The tensile stretching (defocus) ΔL is then:

$$\Delta L = \frac{PL}{AE} = \frac{44.5 \text{ N} \times 1 \text{ m}}{0.001574 \text{ m}^2 \times 69 \text{ GPa}} = 0.4 \text{ μm}$$

Does this exceed the depth of focus of the lens? In Chapter 2, we saw that the depth of focus $\Delta f = \pm 2\lambda(f/\#)^2$, which is $\pm 2(0.5 \text{ μm})(2)^2 = \pm 4$ μm for a visible wavelength ($\lambda = 0.5$ μm) camera. The tensile stretching is thus much less than the allowable depth of focus, giving an initial camera design that is structurally sufficient.

As a check on the stress, we see that the tensile stress $\sigma_t = F/A = 44.5$ N/0.001574 m² = 28.3 kPa. This stress is much less than the microyield strength of aluminum ($\sigma_{MYS} \approx 140$ MPa), so the design is also OK based on the tensile stress from the lens weight. There are, however, other stresses and deflections to consider; these will be reviewed in Sections 5.2 and 5.3.

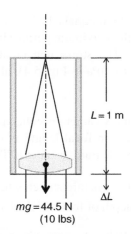

$L = 1$ m

$mg = 44.5$ N
(10 lbs)

ΔL

FIGURE 5.5 Simplified structural model of an aerial camera.

5.2 MECHANICS

Since the tensile strain $\varepsilon = F/AE$, calculating the strain requires knowing how to find the forces on a structure. Knowledge of these forces allows us to estimate the deflections for truss structures, lens and mirror mounts, optical elements, and so on. These deflections determine the tilt, decenter, despace, defocus, and component distortion. When a structure is not moving ("static"), both forces and moments are in equilibrium (i.e., balanced), and the calculation of the forces goes under the title of *mechanics* [5]. In this section, we look at a few examples of how to use the equilibrium concept, to prepare us for the next step in Section 5.3 of finding the deflections under a variety of different stresses.

As a simple illustration of the use of mechanics to find the forces on a structure, Figure 5.6 shows a balance beam of the type that might be used in Olympic gymnastics competitions. The top figure shows the weight of a gymnast (F_o) centered on the middle of the beam, in which case we know intuitively that the weight is equally distributed as $F_o/2$ over each support point. The middle figure shows the extreme case where the gymnast is directly over the support point, in which case her entire weight is supported by it.[1] Finally, the bottom figure shows an arbitrary location for the gymnast, in which case our intuition might need a little help in determining the reaction forces—help supplied by the equations for the forces and moments in equilibrium.

---

[1] The support force is only approximately equal to the gymnast's weight because the beam also has weight, so the full force supplied by the floor through the support is the weight of the gymnast plus the weight of the beam.

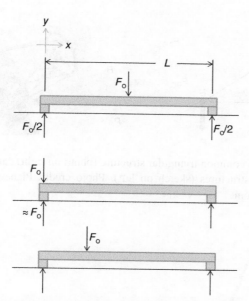

FIGURE 5.6 A balance beam is used to illustrate the principles of force and moment equilibrium to determine the forces on—and deflection of—the structure.

For the xy-coordinate system shown in Figure 5.6c, these equations show that the sum of the forces in the x-direction, the sum of the forces in the y-direction, and the sum of the moments—a force times a distance, measured in units of N-m, and also known as a *torque* or *couple*—in the xy-plane must all be equal to zero for equilibrium balancing [5]

$$\sum F_x = 0 \Rightarrow F_x = 0 \tag{5.2}$$

$$\sum F_y = 0 \Rightarrow F_1 + F_2 - F_o = 0 \tag{5.3}$$

$$\sum M_{xy} = 0 \Rightarrow F_2 L - F_o x = 0 \tag{5.4}$$

The last two equations have two unknowns (reaction forces F_1 and F_2), and can easily be solved to find both. For F_1, the result is $F_1 = F_o(1 - x/L)$, so $F_1 = F_o$ when the load F_o is directly over the left support ($x=0$). This is consistent with our intuition, as is the result that $F_1 = 0$ when the load F_o is directly over the right support ($x=L$). For intermediate cases, the reaction F_1 decreases linearly with distance x as the load F_o moves from the left support to the right.

Looking next at more complex structures typical of optical systems, we can break down the common triangular structure into a collection of trusses (Fig. 5.7).

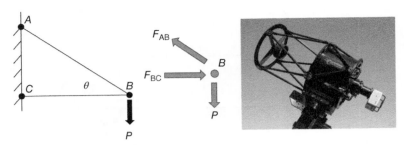

FIGURE 5.7 The common triangular structure (photo on right) can be broken down into simple truss structures (sketch on left). Photo credit: Planewave Instruments, www.planewave.com.

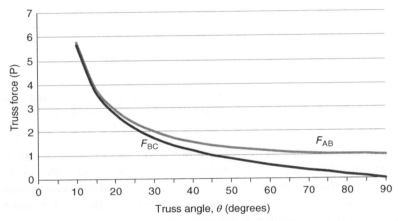

FIGURE 5.8 The force acting on each element of a truss decreases as the truss angle increases, but the length and weight of beam AB increases if the length of beam BC is not changed.

In this case, the center sketch in Figure 5.7 shows that pulling down on the triangle with a load P stretches the beam AB, and compresses the beam BC. Equilibrium is maintained when $\Sigma F_x = 0$ and $\Sigma F_y = 0$; the condition on the x-forces gives $F_{BC} - F_{AB}\cos\theta = 0$, while the condition on the y-forces gives $F_{AB}\sin\theta - P = 0$. Combining these, we find that the tensile force stretching beam AB is $F_{AB} = P/\sin\theta$, and the compressive force on beam BC is $F_{BC} = F_{AB}\cos\theta = P/\tan\theta$.

The point of the calculation is not to master proficiency with solving multiple equations for multiple unknowns, but to understand the important physical trends. These are illustrated in Figure 5.8, where we plot how the force on each beam varies with truss angle θ. Intuitively, we expect that as the angle θ increases—approaching 90° in the extreme case when beam AB is almost vertical—beam AB will carry all of the load, and beam BC will support

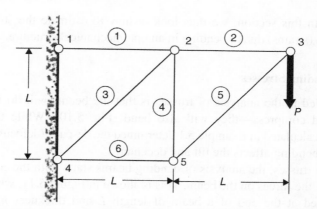

FIGURE 5.9 A longer truss can be composed of smaller elements, but a larger deflection of the end will result. Credit: Prof. David Roylance, MIT OpenCourseWare 3.11 Course Notes.

very little. As designers of the structure, then, this simple example illustrates a very important trade as we change θ: (i) the extra length (and weight) required for beam AB as the angle θ increases; and (ii) larger-area beams—also heavier—are required to resist the bigger forces and higher stresses for smaller angles.

Additional structural-design trends can be illustrated with an even more complex structure, one that builds up on the simple triangular truss with the use of multiple triangles. This is shown in Figure 5.9, where we now have three triangles supporting a load P over a length $2L$ and a height L. In this case, a simple moment equilibrium analysis demonstrates the trend, namely, that a short, stubby beam has smaller reaction forces than a long, thin one. That is, the reaction force that must be provided by the wall depends on the moment applied to the structure at Point "3" (equal to $P \times 2L$) and the moment that the wall supplies to resist (equal to $F_4 \times L$). Moment equilibrium about Point "4" thus shows that the reaction force $F_4 = P \times 2L/L$, or the load P multiplied by the ratio of the structure's length ($2L$) to height (L). This is a very useful result, and will be looked at in more detail in Section 5.3.

5.3   BEAM STRESSES AND STRAINS

While we are now in a position where we can calculate the forces acting on each beam in a structure such as a truss, we do not yet know what the stress and resulting deflections will be. Determining the deflections is the ultimate goal of this chapter, as they affect the tilt, decenter, despace, and defocus of the optical elements. In this section, we thus look at how to estimate the stresses and deflections (strains) due to bending in an optomechanical structure.

5.3.1 Bending Stresses

Not included in the analysis of trusses is that the beams will do more than stretch and compress—they will also bend (Fig. 5.10). While the tensile deflection calculated in Example 5.1 determined the despace (defocus) between elements, bending affects the tilt and decenter.

As with trusses, the analysis of bending beams starts with the equilibrium analysis of the forces on the beam. This is shown in Figure 5.11, where a load F is applied at the end of a beam of length L and thickness h. Moment equilibrium about lower left-hand corner gives $F \times L = F_R \times h$, or $F_R = F \times (L/h)$. We again see that a short, stubby "structure" with small L/h—a beam in this case—has a smaller reaction force F_R than a long, wobbly one—a result that has many implications.

The first of these is that the reaction force F_R depends on the leverage applied to the beam. That is, the beam thickness h resists the *bending moment* (or *torque* or *couple*) $M = F \times L$ applied at the end of the beam. As shown in Figure 5.11, the horizontal reaction forces along the beam thus vary from tensile (F_R pointing to the left) to compressive (F_R pointing to the right) as we move from top to bottom of the cantilevered beam; there is also a *neutral axis* where the reaction force is

FIGURE 5.10 Bending of a beam affects the tilt and decenter of an optomechanical structure.

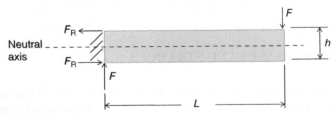

FIGURE 5.11 A structural geometry known as a cantilever beam is supported on one side and has a force F applied at any point along the length of the beam.

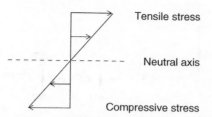

Tensile stress

Neutral axis

Compressive stress

FIGURE 5.12 The reaction stresses for the beam loading shown in Figure 5.11 vary linearly from tensile to compressive as we move from top to bottom of the beam.

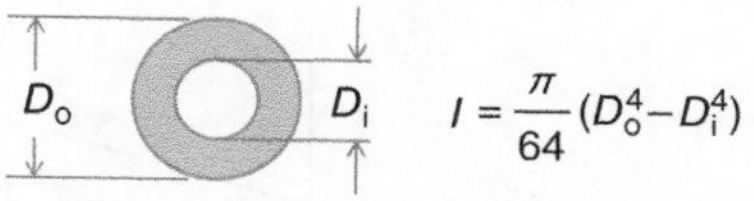

$$I = \frac{\pi}{64}(D_o^4 - D_i^4)$$

D_o D_i

FIGURE 5.13 The geometric resistance to bending increases with the fourth power of the tube outer diameter D_o.

zero (Fig. 5.12). Because the forces are tensile and compressive, we expect the strain in the beam to depend on the elastic modulus E.

For a symmetric beam, the distance y from the neutral axis gives the maximum stress at the top and bottom of the beam where $y = \pm h/2$. The bending stress σ_b in a beam depends on the bending moment M, the distance y from the neutral axis, and a geometric resistance to bending I

$$\sigma_b = \frac{My}{I} \tag{5.5}$$

The geometric resistance to bending—or bending moment of inertia I (units of m^4)—depends on the beam dimensions [2].[2] For a circular tube typical of optomechanical structures, it increases with the fourth power of the diameter (Fig. 5.13)—reducing the bending stress by D^3.

The bending moment of inertia I is also called the *area moment* or *second moment* (Fig. 5.14). The second moment is much more effective at resisting bending when y^2 is large, resulting in $I \sim h^3$ or D^4. The tabulation of the second moment for a variety of different cross-sectional geometries is available in any standard textbook on mechanical design [6, 7].

Because of the dependence of the bending resistance I on distance from the neutral axis, most of the material can be removed from a rectangular beam without significantly reducing its stiffness. That is, an "efficient" structure can

[2] The bending moment of inertia I is a different concept from the rotational moment of inertia for rigid-body motion. See Ref. [5] for more details.

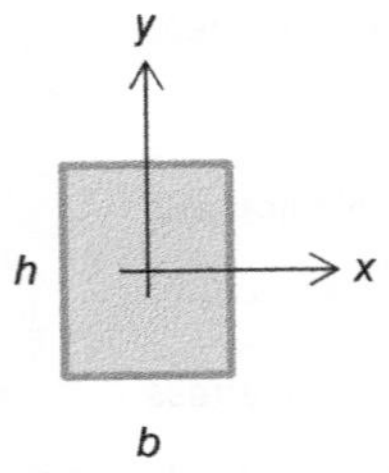

$$I = \int_{-h/2}^{+h/2} y^2 dA = b \int_{-h/2}^{+h/2} y^2 dy = \frac{bh^3}{12}$$

FIGURE 5.14 A rectangular beam illustrates the idea of the beam height h contributing to the bending resistance I via the second moment y^2.

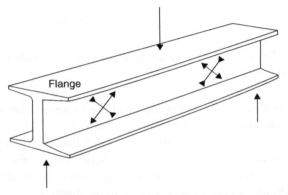

FIGURE 5.15 The I-beam efficiently resists bending via the wide flanges at the top and bottom, the furthest distance from the neutral axis. Adapted from J. E. Gordon, *Structures: Or Why Things Don't Fall Down*, DaCapo Press (2003).

be designed—that is, one with maximum stiffness using minimum weight and material—by placing as much material as possible as far away as possible from the neutral axis. This is the idea behind the I-beam, so named because its shape is the capital English letter "I" (Fig. 5.15). This concept is also used for the hollow tube shown in Figure 5.13—a more efficient structure than a solid bar. This concept of structural efficiency will be looked at in more detail in Section 5.4.

Example 5.2   We saw in Example 5.1 how to calculate the tensile deflection of a camera tube under vertical gravity loading of a lens. In this example, we look at the bending stresses of the same camera under wind loading (Fig. 5.16). What is the stress in the lens tube with a wind force which puts a distributed load $F_w = 100\,\text{N}$ (22.5 lb) on the camera, and is this acceptable?

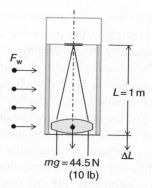

F_w

$L = 1\,m$

ΔL

$mg = 44.5\,N$
(10 lb)

FIGURE 5.16 The aerial camera of Example 5.1 will be bent due to a wind force F_w.

Using Equation 5.5, the maximum bending stress—at the base of the tube where the moment is largest—is

$$\sigma_b = \frac{My}{I} = \frac{0.5 F_w L\,(D/2)}{\pi\,(D_o^4 - D_i^4)/64} = 0.25\,\text{MPa (36.3 psi)}$$

This result assumes the wind force F_w is uniformly distributed over the length of the camera, giving an average location of $L/2$ along the length of the tube where the wind is applied. In addition, the dimension $y = D_o/2$ and the bending resistance I have been calculated from the tube dimensions given in Example 5.1. The resulting stress is very low—much less than the microyield stress, even after adding the stress from the tensile loading of the lens. What is not yet clear, however, is whether or not the *deflection* is acceptable. This is reviewed in the next section.

5.3.2 Bending Strain

The initial alignments established at assembly for tilt, decenter, despace, and defocus may not be maintained when loads are applied to the system. All structures deflect when loads are applied; these deflections may be due to rigid-body ("rocking") deflection or may occur because the structure itself bends or twists. A flexible structure allows a large degree of relative motion between the elements, resulting in both image motion and an increase in blur size (Chapter 4). The degree of flexibility that's acceptable flows down from the system requirements—these determine the structural materials, geometry, and need for alignment mechanisms or rotationally insensitive optics such as pentaprisms.

Given the stress in a bending beam, it is straightforward to determine the strain. In Section 5.1, we saw that stress and strain are related by the material stiffness (elastic modulus E) for linear strains ($\varepsilon = \sigma/E$). Since the bending stress is given by $\sigma_b = My/I$ (Eq. 5.5), the strain over the elastic range is given by $\varepsilon = \sigma_b/E = My/EI$. The quantity in the denominator of this equation—the EI-product—is called the *bending stiffness* or *flexural stiffness*, a combination of material and geometric properties needed to characterize bending strain.[3]

This is the strain in each atomic layer of the beam, however, and does not indicate how much the end of the beam will tilt or decenter due to an applied load at its end. With a beam stretched in tension, each layer parallel to the load is deflected; with bending, each layer perpendicular to the load is stretched or compressed—and by a different amount, depending on where the layer is located with respect to the neutral axis (Fig. 5.17).

A simple tube structure is shown in Figure 5.18, which illustrates that a downward load—due to self-weight and any additional applied force F—creates tilt and decenter of the structure with respect to its fixed end. The tilt and decenter depend on the load, the material, and the shape and size of the structure. For a cantilevered beam that is clamped on one end and has a lens weight or other force that is much larger than the tube weight on the other end, the decenter (or deflection) δ is given by [2, 6]

$$\delta = -\frac{FL^3}{3EI} \tag{5.6}$$

This equation is valid when a heavy lens is located at the end of a much lighter tube. Although this is not the case for many designs of interest, the equation is not much different for cases that are of interest and serves to illustrate the physical principles. These principles are that the resistance to decenter has both a material and geometric component. The material resistance is the elastic modulus E. The geometric component has a much larger impact on the deflection through the length L and the bending moment of inertia I. Given the D^4-dependence of I,

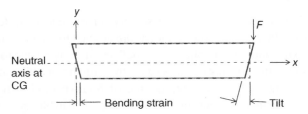

FIGURE 5.17 Strain in a cantilevered beam increases as we move away from the neutral axis, in a manner that compresses the bottom of the beam and stretches the top when the load is applied as shown.

---

[3] In comparison, determining the tensile strain only requires the elastic modulus E, not the EI-product.

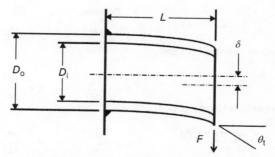

FIGURE 5.18 A cantilevered tube deflects when pushed or pulled by a force F; hence, the centerline of the tube on one end is decentered by an amount δ with respect to the centerline on the clamped end. The centerline is also tilted (not shown) by an angle θ_t.

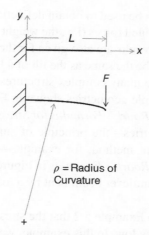

FIGURE 5.19 The radius of curvature ρ of a cantilevered beam varies along its length, and depends on the force F through the relation $1/\rho = F(x-L)/EI$.

the deflection thus depends on the ratio of the beam's length to its height, as do the reaction forces holding the beam in place (see Fig. 5.11).

Equation 5.6 can be obtained by looking at the beam as bending with a radius of curvature ρ that varies along its length, and thus depends on the coordinate x (Fig. 5.19). For a beam that is long compared with its thickness, elementary calculus shows that this radius depends on the second derivative of the shape $y(x)$, where $1/\rho \approx d^2y/dx^2$. Physically, we also know that the radius gets smaller as the moment M $[=F(x-L)]$ increases and larger as the flexural stiffness EI increases, such that $1/\rho = M/EI$. Equating the mathematical and physical pictures and integrating the result gives us Equation 5.6 for the case of a load F on the end of the cantilever [2].

	Decenter	Tilt
 Bending moment M at end	$\delta = -\dfrac{ML^2}{2EI}$	$\theta_t = \dfrac{ML}{EI}$
Concentrated load F at end	$\delta = -\dfrac{FL^3}{3EI}$	$\theta_t = \dfrac{FL^2}{2EI}$
 Distributed load $w = F/L$	$\delta = -\dfrac{wL^4}{8EI}$	$\theta_t = \dfrac{wL^3}{6EI}$

FIGURE 5.20 Decenter and tilt for a cantilevered beam subject to various loading conditions [2].

The same principles can be used to obtain deflection results for beams subject to moments and distributed forces (i.e., the weight of a beam). These results are shown in Figure 5.20; results are also given for the tilt angle of the end face of the beam, which would be the same as the tilt of a lens attached to this face.

Unfortunately, there are many complex structures and loads which cannot be analyzed using the simple geometries shown in Figure 5.20. The standard reference for such cases is *Roark's Formulas for Stress and Strain* [7]. Methods for handling these geometries—the principle of superposition, parallel-axis theorem, and area-moment method, for example—are also covered in this reference. The table from *Roark's Formulas* in Figure 5.21 shows how to analyze a general case of a cantilever beam bent by a load.

Example 5.3 We saw in Example 5.2 that the stress at the base of a cantilevered lens tube can be very low. In this example, we approximate the tilt and decenter of the lens mounted in the end of this tube (see Fig. 5.16).

For the same wind forces as Example 5.2—uniformly distributed over the length of the lens tube—we find that the force per unit length $w = F_w/L$. For this case, the tilt and decenter equations from Figure 5.20 are $\theta_t = wL^3/6EI = F_w L^2/6EI$ and $\delta = wL^4/8EI = F_w L^3/8EI$. Substituting for $F_w = 100\,\mathrm{N}$, $L = 1\,\mathrm{m}$, $E = 69 \times 10^9\,\mathrm{N/m^2}$, and $I = 4.94 \times 10^{-5}\,\mathrm{m^4}$, we find that $\theta_t = 4.9\,\mu\mathrm{rad}$ and $\delta = 3.7\,\mu\mathrm{m}$.

These results are approximate since an assumption used in the derivation of the tilt and decenter equations is that the beam is long compared with its thickness ($L > 5D$). The length-to-diameter ratio in this case does not meet this criterion, so the results obtained are not in any way accurate, but do give us a starting point for comparing different designs. In this case, the design shows that tilt and decenter that may be significant, and thus requires a more refined analysis. The example also illustrates that while the stress in an optomechanical system is likely to be low, the deflections such as tilt and decenter may not be.

1. Concentrated intermediate load

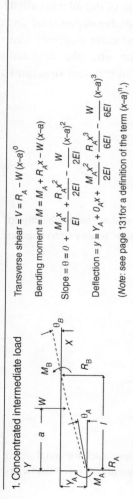

Transverse shear $= V = R_A - W(x-a)^0$

Bending moment $= M = M_A + R_A x - W(x-a)$

Slope $= \theta = 0 + \dfrac{M_A x}{EI} + \dfrac{R_A x^2}{2EI} - \dfrac{W}{2EI}(x-a)^2$

Deflection $= y = Y_A + 0_A x + \dfrac{M_A x^2}{2EI} + \dfrac{R_A x^3}{6EI} - \dfrac{W}{6EI}(x-a)^3$

(Note: see page 131 for a definition of the term $(x-a)^n$.)

End restraints, reference no.	Boundary values	Selected maximum values of moments and deformations
(1a). Left end free, right end fixed (cantilever)	$R_A = 0$ $M_A = 0$ $\theta_A = \dfrac{W(l-a)^2}{2EI}$ $y_A = \dfrac{-W}{6EI}(2l^3 - 3l^2 a + a^3)$ $R_B = W$ $M_B = -W(l-a)$ $\theta_B = 0$ $y_B = 0$	Max $M = M_B$: max possible value $= -Wl$ when $a = 0$ Max $\theta = \theta_A$: max possible value $= \dfrac{Wl^2}{2EI}$ when $a = 0$ Max $y = y_A$: max possible value $= \dfrac{-Wl^3}{3EI}$ when $a = 0$
(1b). Left end guided, right end fixed	$R_A = 0$ $M_A = \dfrac{W(l-a)^2}{2l}$ $\theta_A = 0$ $y_A = \dfrac{-W}{12EI}(l-a)^2(l+a^2)$ $R_B = W$ $M_B = \dfrac{-W(l^2 - 2a)}{2l}$ $\theta_B = 0$ $y_B = 0$	Max $+M = M_A$: max possible value $= \dfrac{Wl}{2}$ when $a = 0$ Max $-M = M_B$: max possible value $= \dfrac{-Wl}{2}$ when $a = 0$ Max $y = y_A$: max possible value $= \dfrac{-Wl^3}{12EI}$ when $a = 0$

FIGURE 5.21 Table from a typical design handbook for determining the deflection of complex beam geometries and loads. Adapted from Warren C. Young and Richard G. Budynas, *Roark's Formulas for Stress and Strain*, McGraw-Hill (2001).

5.3.3 Shear Stresses and Strains

What happens to a truss when the diagonal element (i.e., beam 3 in Fig. 5.15) is removed? In this case, the truss looks like an open box structure, which we know from experience folds fairly easily (Fig. 5.22). The reason is that the diagonal provides tensile and compressive reaction forces to balance the vertical force—a force that bends the truss, but also tends to shear the structure.

A shear force is one that forces atomic planes to slide over each other, rather than pull away from each other (as in tension). For example, Figure 5.23 shows that a tube will twist with a load that is not centered, with the resistance to twisting determined in part by a material property known as the shear modulus G (typically smaller than the elastic modulus E by a factor of 2.5 or so). This rotational sliding of different planes along the length of the tube is called shear, provided by a shear force V. The amount of twist also depends on the geometry of the tube, with Figure 5.23 showing an efficient structure that places material a large distance from the center of the tube (the torsional neutral axis). The next section looks at more examples of efficient structures used in optomechanical design.

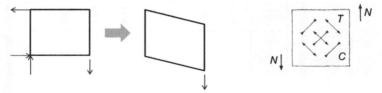

FIGURE 5.22 An open box structure has no resistance against shear loads; the closed box on the right—such as a solid material—forces the diagonal "beams" into tension T and compression C.

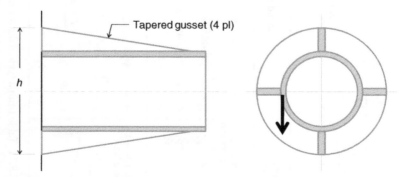

FIGURE 5.23 An asymmetric, offset force will twist a structure; such torsional strain can be reduced with tapered gussets that provide a larger dimension to resist the twisting, without excessively increasing weight.

5.4   STRUCTURAL GEOMETRIES

Summarizing the chapter up to this point: A beam resists bending mostly due to material at the extremes, such that thicker beams have less deflection ($\delta = FL^3/3EI \sim 1/h^3$ for a rectangular beam). The beam can be made more "efficient"—that is, the same stiffness with less weight, material, and cost—by removing material which is not effective at resisting deflection. I-beams and cylindrical tubes are thus common mechanical components for structural design (Fig. 5.24).

We also know that along the length of the beam, the stresses are largest at the support where the moment and shear force are largest (Fig. 5.25a). The beam can thus be made even more efficient by removing the material at the end which is not doing much to support the load (Fig. 5.25b—see also Fig. 5.23 for the use of a taper in the gussets used to reduce torsional twisting).

Combining the I-beam concept with the lengthwise taper, we can make a structure even more efficient by removing material from the center of the

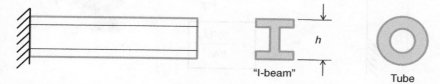

FIGURE 5.24 I-beams and cylindrical tubes are common mechanical components for structural design due to their efficiency (stiffness per unit weight, cost, etc.).

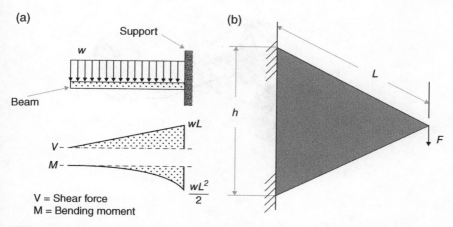

FIGURE 5.25 With the shear force and bending moment largest at the support (a), beams can be made more efficient with a taper to remove the material at the end which provides little resistance against deflection (b).

tapered section which does little to support the load. This results in the triangular structure shown in Figure 5.26, with the end deflection δ decreasing as the base height h and the cross-sectional area A are made larger. The dimension h thus acts to resist bending, not just the area of the beam itself; this is a huge structural advantage, and the reason why optomechanical structures look the way they do. Another example is shown in Figure 5.27, where the "spider" structure holding the secondary mirror also uses triangles (where one leg of the "triangle" is curved).

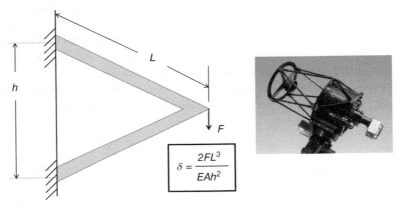

$$\delta = \frac{2FL^3}{EAh^2}$$

FIGURE 5.26 Combining the I-beam concept with the lengthwise taper results in the triangular shape commonly used in efficient optomechanical structures. Photo credit: Planewave Instruments (www.planewave.com).

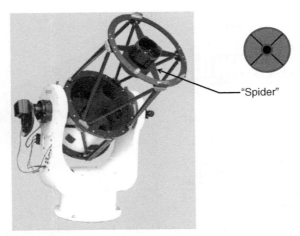

FIGURE 5.27 Triangular structures are used in many commercial telescope designs to reduce weight and cost. Photo credit: Planewave Instruments (www.planewave.com).

Another common structural geometry used to maintain optical alignment with maximum stiffness and minimum weight is the optical bench on which an instrument is assembled (Fig. 5.28). These can range from simple aluminum plates to honeycomb graphite-epoxy (GrE, also known as "carbon fiber") structures to meet a range of requirements pertaining to stiffness, weight, flatness, jitter, and long-term stability (drift). Honeycomb benches resist deflection through their thickness d, and maintain low weight by using a honeycomb core with high stiffness-to-weight ratio. An off-the-shelf honeycomb bench with stainless steel facesheets is shown in Figure 5.28. While useful for terrestrial systems, the stainless steel is too heavy for space-based instruments; instead, GrE—or any of the other graphite-based materials such as graphite-cyanate—are commonly used even though they have their own design details (e.g., "outgassing" and water loss or "desorption" that alters the flatness).

This same honeycomb geometry is also used for the design of large glass mirrors (Fig. 5.29). Typically the hexagonal shape of the glass honeycomb structure can be replaced with rectangular or triangular patterns without compromising stiffness or paying a large penalty in weight. With the bonding of front

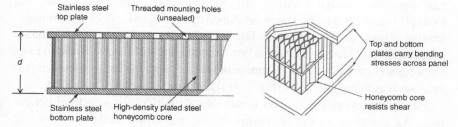

FIGURE 5.28 A common optical bench resists deflection via its thickness d and maintains low weight by using a hexagonal, honeycomb core. Credit: CVI Laser, LLC.

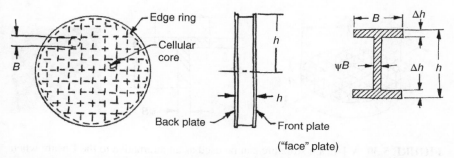

FIGURE 5.29 The I-beam concept can also be used in the structural design of glass mirrors. Credit: Dan Vukobratovich, *Introduction to Optomechanical Design*, SPIE Short Course SC014.

and back plates—as in the honeycomb bench shown in Figure 5.28—the geometry relies on the I-beam principle to maximize stiffness and minimize weight. A difficulty related to the fabrication of such mirrors is that the top faceplate—which is coated with a thin layer of aluminum or gold to create a mirror—cannot be too thin, or the faceplate will reflect the underlying honeycomb structure. This process is known as an "imprinting" or "quilting" of the faceplate, and reduces mirror quality in the same way a "funhouse" mirror distorts images.

Large mirrors have also been fabricated using a similar structural geometry known as a T-beam. An example of a triangular-core mirror fabricated from silicon carbide (SiC) is shown in Figure 5.30. The T-beam was used because of the low loading—and low weight—of a mirror to be used in space. The advantages and disadvantages of SiC as a structural material will be examined in more detail in Section 5.5 and Chapter 6.

While the focus of this section has been on efficient structural geometries, it is not necessary for every design to meet this requirement. For example, the "box" structure is often used when shape and packing efficiency are more important than peak structural efficiency—electronics boxes for rectangular electronics boards are a typical example. Even boxes, however, have details that can make their design relatively stiff in comparison with their weight. An example is a space-qualified laser developed for a laser-radar system for remote-sensing of atmospheric pollutants (Fig. 5.31). While weight for this space-based instrument was clearly important, a box structure had sufficient stiffness to meet the weight requirement. This requirement was met with the judicious use of features such as lightening holes (Fig. 5.31a) and structural ribs (Fig. 5.31b).

Finally, the so-called support conditions also affect the structural deflection. As a simple example, compare the maximum deflection for the supports shown in Figure 5.32. The support on the left of Figure 5.32 is the conventional cantilever beam we have looked at in detail in this chapter.

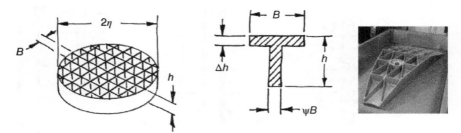

FIGURE 5.30 A T-beam structure can be used as an alternative to the I-beam when the loading requirements are lower. Photo credit: European Space Agency. Illustration credits: Dan Vukobratovich, *Introduction to Optomechanical Design*, SPIE Short Course SC014.

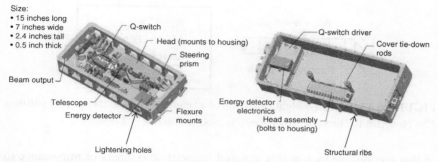

FIGURE 5.31 Simple box structures can be made more efficient with the use of lightening holes and structural ribs. Credit: Floyd Hovis, Proc. SPIE, Vol. 6100 (2006).

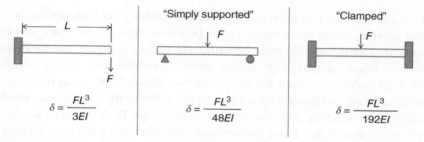

FIGURE 5.32 The "support" or "end" conditions of a beam also determine its deflection.

The middle illustration shows the effects of holding the beam on both ends in a manner known as "simply supported" (i.e., not clamped); the center deflection is a factor of 16× smaller than the end deflection of a cantilever, of which a factor of $L^3 = 2^3 = 8×$ is due to the shorter length over which the beam is not supported. Finally, if the simply-supported beam is now clamped, the deflection is now a factor of 4× smaller still.

5.5 STRUCTURAL MATERIALS

We have seen that in a number of structural geometries it is important that the stiffness of a structure be maintained—to maintain tilt, decenter, despace, and defocus—with minimum weight. At the same time, the materials used must themselves have this same characteristic, a material performance metric summarized by the stiffness-to-density ratio, or *specific stiffness* E/ρ. That is, in designing a structure, the materials should not be so heavy compared with their stiffness that they deflect excessively under their own weight (Fig. 5.33).

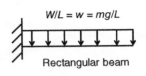

$$\delta = \frac{WL^3}{8EI} = \frac{3WL^3}{2Ebh^3}$$
$$= \frac{3mgL^3}{2EAh^2} = \frac{3gL^4}{2h^2}\left(\frac{\rho}{E}\right)$$

W/L = w = mg/L

Rectangular beam

FIGURE 5.33 The self-weight deflection of a beam depends on the specific stiffness E/ρ. Note that the mass density $\rho = m/V$ is used, not the weight density.[4]

In this section, we look at this material property, and others of importance to structural design as well.

5.5.1 Specific Stiffness

A typical application of the specific stiffness is the design of high-performance scan mirrors, where the mirror (i) must be extremely light, so that it can be scanned quickly; and (ii) must be extremely stiff to maintain optical shape (surface figure) during large accelerations. Beryllium and silicon carbide (SiC) are excellent options for such mirrors—although SiC is an inherently brittle ceramic, and thought by some to be risky. Beryllium, on the other hand, is difficult to machine due to its toxicity to the lungs. Both materials are also very expensive. While lower performance, graphite-epoxy (GrE) is a lower-cost alternative to Be and SiC, but cannot at this time be polished into a mirror. It is lighter and stiffer than many common materials (aluminum, e.g.), leading to its use even in high-end commercial telescopes (see photo in Fig. 5.26).

In addition to scan mirrors, the specific stiffness is also used as a metric for comparing different structural materials, but in practice it must be used with caution, as loads larger than the self-weight reduce the benefits of large E/ρ. A comparison of a number of common structural materials is shown in Figure 5.34, where the elastic modulus is plotted against the density. The dashed line shows materials with the same E/ρ, the most notable feature of which is that many materials have the same specific stiffness, including aluminum, titanium, magnesium, and stainless steels [8]. Also shown are the extremely stiff materials such as SiC—which is slightly heavier than aluminum—and beryllium, with the highest-known E/ρ.

It is not correct to conclude from Figure 5.34, however, that aluminum and steel are equallyviable options for optomechanical structures. Figure 5.35

---

[4] In the definition of mass density ρ, the mass m (units of kg) is divided by the volume V (m³), giving $\rho = m/V$. It is unfortunate that the kilogram is also commonly used as a unit of weight. The actual unit of weight in the MKS system is the newton (N); it is equal to the mass multiplied by the acceleration due to gravity ($F = mg$), or kg×9.81 m/s², such that 1 lb = 0.454 kg×9.81 m/s² = 4.45 N. While the use of "kilogram" for weight is fine for grocery stores, it leads to serious problems in engineering.

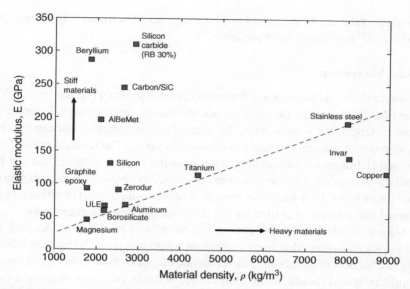

FIGURE 5.34 A scatter plot showing the elastic modulus and material density for various structural materials. Credit: Keith B. Doyle, Victor L. Genberg, and Gregory J. Michels, *Integrated Optomechanical Analysis* (2nd ed.), SPIE Press (2012).

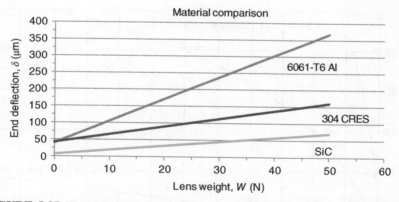

FIGURE 5.35 The end deflection of a cantilevered lens tube depends on both the self-weight deflection and the lens weight that must be supported.

shows why, where the end deflection of a cantilevered cylindrical tube with a lens mounted at its end has been plotted against lens weight. At zero load (no lens), the self-weight deflection of aluminum (6061-T6) and stainless steel (304 CRES) are about the same ($\delta \approx 40\,\mu m$ for $D_o = 100$ mm, $D_i = 96$ mm, and $L = 1$ m). As the lens weight increases, however, the aluminum deflects more—at a rate of 3× that of steel, as E now plays more of a role in the total deflection than E/ρ. Steel, however, is ~3× heavier, and the diameter of the aluminum tube can be increased to give it equivalent stiffness at lower weight (since $I \sim D^4$ for tubes—see Problem 5.9).

5.5.2 Microcreep

Beyond a threshold stress, some materials continue to strain ("creep" or "drift") over time, even though the stress is not increasing. Such strain is inherent in some materials, but is made worse by internal residual stresses and can thus be alleviated with thermal annealing or cryogenic cycling. The residual stress from external (fabrication) sources also determines the creep over time, and can be similarly alleviated. Micron-scale creep—or microcreep—can be particularly troublesome for high-precision optical systems which must maintain tight dimensional stability over time for tilt, decenter, and despace alignments.

Measured microcreep data [9]—given in units of microinches per inch or μm per meter—show that some aluminum alloys creep much less than others (Fig. 5.36). Soft materials such as RTVs and epoxies, on the other hand, creep significantly and should not be used for heavy loads, precision alignments, or high temperatures. Microcreep data can be extrapolated to predict strain *for the same material* in terms of the applied stress, typically using an empirical power law such as Andrade's, where the time-varying strain $\varepsilon(t) = \beta t^n$ for a

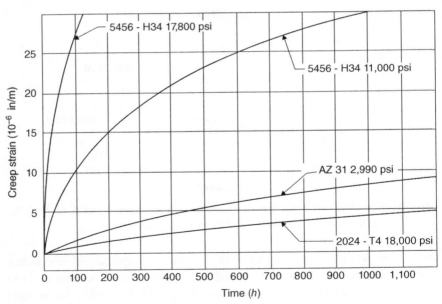

FIGURE 5.36 Strain may increase with time, even though the load and stress are constant. Adapted from C. W. Marschall and R. E. Maringer, *Dimensional Instability*, Pergamon Press (1977).

stress-dependent parameter β and exponent $n \sim 1/3$ [9]. Unfortunately, dimensional stability information such as that shown in Figure 5.36 is difficult to find, and it is often necessary to remove residual stresses as best as possible, and measure the microcreep to verify that it is sufficiently small.

5.5.3 Materials Selection

The selection of an appropriate structural material entails more than the specific stiffness and dimensional stability. For example, we have also seen that the total weight (via density ρ), microyield stress σ_{MYS}, or elastic modulus E may determine material selection. Table 5.1 lists the values of these properties for a number of commonly used structural materials. Note that these are only *structural* properties, and there will likely be other materials properties which must be considered—vibration (Chapter 7), thermal (Chapter 8) or kinematic (Chapter 9)—and combined for a system-level comparison of optomechanical performance (Chapter 10).

Summarizing: Forces and moments ("loads") act on any optical system, and may be due to self-weight and applied forces of various types—acceleration, shock, vibration, wind, water pressure, and so on. These loads cause the optical structure to stretch, compress, bend, and twist. The degree of stretching, bending, and so on depends on the structural efficiency through both geometry (cross-sectional area and bending moment of inertia) and material properties (elastic modulus and specific stiffness). The structural deflections result in tilt, decenter, despace, and defocus of the optical elements.

TABLE 5.1 Mechanical Properties of Typical Structural Materials

Material	Density, ρ (kg/m^3)	Elastic modulus, E (GPa)	Specific stiffness, E/ρ (MN-m/kg)	Microyield strength, σ_{MYS} (MPa)
Aluminum, 6061-T6	2700	69	26	140
Beryllium, S-200FH	1850	303	164	40
Brass, 70–30	8530	110	13	160
Copper, OFHC	8940	115	13	12
Graphite epoxy	1800	95	53	–
Invar 36	8050	141	18	20
Magnesium, AZ-31	1770	45	25	33
Silicon carbide, RB	2900	320	110	–
Stainless steel, 304	7860	193	25	50
Stainless steel, 440C	7700	200	26	480
Titanium, 6Al-4V	4430	114	26	50

Data are from Vukobratovich [3] and Yoder [4].

Optical performance is thus directly affected by the structural design. In the next chapter, we look at how forces and moments also distort—and possibly break—the optical elements themselves.

PROBLEMS

5.1   A spring gets longer as the force F increases and the spring stiffness k decreases, such that $F = k\delta x$, where δx is the deflection (change in length). For a given force F, what is the total deflection for two different springs ($k_1 \neq k_2$) in series? If we use springs with the same stiffness, what is the connection between the change in length and the structural property of strain, where a longer beam in tension extends more than a shorter one?

$$F \longleftarrow \text{---}\bigwedge\!\bigwedge\!\bigwedge\!\!\text{---o---}\bigwedge\!\bigwedge\!\bigwedge\!\text{---} \longrightarrow F$$
$$k_1 \qquad\qquad k_2$$

5.2   For a given force F, what is the total deflection for two different springs in parallel (i.e., their deflection is the same)? If we use springs with the same stiffness, what is the connection between the force on the springs and the structural property known as stress, where a thick beam in tension has less stress than a thin one?

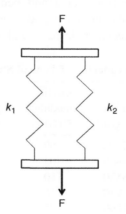

5.3   Quantitatively, how does the bending stress of a beam vary with beam thickness or height h? How does this compare with the deflection δ?

5.4   With the wind loading of an aerial camera as shown in Figure 5.16, which side of the tube will have the lowest stress—the side directly exposed to the wind, or the far side? Or is there no difference? Hint: the tensile stresses from the weight of the lens must be included.

5.5 In Figure 5.19, we showed that the length-dependent radius of curvature of a cantilevered beam depends on the force F through the relation $1/\rho = F(x - L)/EI$. Given this, what is ρ at the end of the beam? At the base of the beam where is it attached?

5.6 The decenter due to deflection for the end-loaded geometry shown in Figure 5.20 is given by $\delta = FL^3/3EI$. Why isn't the decenter equal to the tilt angle multiplied by the length?

5.7 When pulled in tension, what is the tensile spring constant k_T of a rectangular beam in terms of its elastic modulus and geometry? When bent by a concentrated load F on its end, what is the bending spring constant k_{bend}?

5.8 What is the self-weight deflection of an aluminum cylinder? Assume the cylinder length $L = 1000\,mm$ and the outside diameter $D_o = 100\,mm$. How does it compare with that of a hollow tube with the same length and outside diameter, but with an inside diameter $D_i = 96\,mm$?

5.9 In Section 5.5.1, it was mentioned that an aluminum lens tube can be redesigned with a larger diameter to give it the same stiffness and deflection as a steel lens tube. Is the aluminum lens tube heavier or lighter than the stainless steel tube? Assume the tube is loaded at its end with a lens whose weight is 50 N. Does the diameter depend on the weight of this lens?

REFERENCES

1. J. E. Gordon, *Structures: Or Why Things Don't Fall Down*, New York: DaCapo Press (2003).

2. J. P. Den Hartog, *Strength of Materials*, New York: Dover Publications (1949).

3. D. Vukobratovich, *Introduction to Optomechanical Design*, SPIE Short Course (www.spie.org) (2009).

4. P. R. Yoder, Jr., *Opto-Mechanical Systems Design* (3rd Edition), Boca Raton: CRC Press (2005).

5. J. P. Den Hartog, *Mechanics*, New York: Dover Publications (1948).

6. V. M. Faires, *Design of Machine Elements* (4th Edition), New York: Macmillan (1965).

7. W. C. Young and R. G. Budynas, *Roark's Formulas for Stress and Strain*, New York: McGraw-Hill (2001).

8. K. B. Doyle, V. L. Genberg, and G. J. Michels, *Integrated Optomechanical Analysis* (2nd Edition), Bellingham: SPIE Press (2012).

9. C. W. Marschall and R. E. Maringer, *Dimensional Instability*, London/New York: Pergamon Press (1977).

6

STRUCTURAL DESIGN— OPTICAL COMPONENTS

While most of us think of windows as something to look through while keeping the rain and wind out of the house, there are those who view them as a matter of life or death. The astronauts on NASA's Space Shuttle, for example, relied on these brittle components to protect them from the cold vacuum of outer space and the massive heat of atmospheric re-entry. Similarly, bathysphere operators place their lives in the hands of optomechanical engineers who design Plexiglas windows to survive the pressure of the ocean's depths (Fig. 6.1), without fracturing catastrophically into multiple shards.

Before windows reach the point of failure, however, there are equally important—but less catastrophic—consequences of poor window design. The performance of an optical system that looks through a window, for example, depends not only on the deflection of mechanical structures that hold the optics in alignment but also on the distortion of the windows themselves. So just as the structural design of the mechanical components led to the idea of a strain-limited structure—as distinct from strength-limited—the optical components are typically limited in the same manner.

The purpose of this chapter is to review both the strain and strength limitations of window performance. Other optical components are included as well—lenses, mirrors, prisms, and so on—where the strain of the surfaces may change the surface power and surface figure to a degree that is unacceptable.

Optomechanical Systems Engineering, First Edition. Keith J. Kasunic.
© 2015 John Wiley & Sons, Inc. Published 2015 by John Wiley & Sons, Inc.

FIGURE 6.1 The windows in a bathysphere protect the hydronaut from being crushed by the pressure of the ocean's depths. Photo credit: NOAA (http://ocean explorer.noaa.gov).

Such deflections are difficult to analyze and are typically predicted analytically using finite-element analysis (FEA)—see Chapter 10 and Ref. [1] for more details.

This chapter begins with a description of windows, lenses, and so on, from the perspective of structural design (Section 6.1.1); as in the rest of the book, the tools used are back-of-the-envelope and order-of-magnitude estimates. A complication not seen with one-dimensional beams in Chapter 5 is that an optical component that is stretched or squeezed in the radial direction, for example, will also get thinner or thicker—a critical parameter in the performance of an optical component, controlled by a quantity known as Poisson's ratio (Section 6.1.2). The remaining parts of Section 6.1 then review a few of the design concepts and techniques needed to minimize structural strain and wavefront error (WFE), including plate bending (Section 6.1.3), contact stresses (Section 6.1.4), and stress concentrations (Section 6.1.5).

The structural strength of brittle materials such as glasses, semiconductors, and ceramics is covered in Section 6.2. Two approaches are used to analyze their strength: fracture mechanics (Section 6.2.1) and Weibull statistics (Section 6.2.2). Brittle materials are much more sensitive to surface cracks than ductile materials such as aluminum, and a different metric for evaluating strength known as *fracture toughness* is used. Statistical concepts such as characteristic strength are also useful. The chapter closes with a discussion of materials selection from the optomechanical perspective, focusing on the structural deformations that affect WFE.

6.1   STRUCTURAL PLATES

In structural terms, windows, lenses, and mirrors can be thought of as *plates*, that is, a beam in two dimensions and "thin" compared with these dimensions (Fig. 6.2). For example, in the commercial world of architecture and construction, windows are sometimes known as plate-glass windows—a description not of the glass composition but of the window size and thickness. As the thickness starts to approach its size, we then have a block of glass, and the distinction between plates and blocks is based purely on the mathematical tools available to analyze the component. In this section, we review the first-order analysis of plates as optomechanical structures.

6.1.1   Windows, Lenses, and Mirrors

One of the most important requirements for an optical mount is that is does not strain an optical component excessively. The strain in the case of optical components refers to changes in surface curvature and surface figure. Applied forces can create such strain, changing surfaces from nearly spherical with diffraction-limited performance to something highly nonspherical with large WFE. A typical example is that of a lens being clamped around its circumference and expanding in thickness as it is strained. This may be due to a

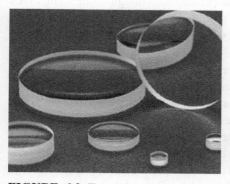

FIGURE 6.2 From the perspective of optomechanical structural design, most windows, lenses, and mirrors are known as plates. Photo credits: CVI Laser, LLC.

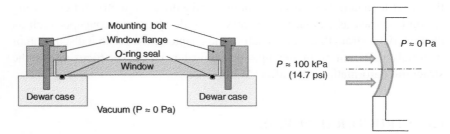

FIGURE 6.3 The pressure difference ΔP across a vacuum or Dewar window forces the window to bow into a weak negative lens. Left illustration adapted from Oli Durney, University of Arizona.

difference in thermal contraction, where the lens mount shrinks faster than the lens when their temperature is changed.

Another example is a vacuum or Dewar window, where an evacuated volume is kept sealed (Fig. 6.3). As with the structural materials reviewed in Chapter 5, optical materials such as glasses, crystals, and ceramics deflect when a force is applied to them. Forces can be due to aerodynamics, water pressure, acceleration, and the like. In the case of Dewar or vacuum windows, the force is an air pressure difference ΔP across the window, giving a force $F = A\Delta P$ (with units $N = m^2 \times Pa$). The force bows the window, so a flat window deflects into a weak negative lens that creates defocus (see Problem 6.1).

If symmetric, this bowing does not create an optical path difference (OPD) or WFE in the sense of surface-figure irregularities such as coma or astigmatism; the defocus may or may not be a problem, depending on its magnitude and whether focus can be adjusted at assembly and the forces (such as $A\Delta P$) do not change. It is possible to accommodate this defocus by adjusting the window-to-focal plane array (FPA) distance before reaching the operating conditions of vacuum and cryogenic temperatures, although doing so involves a lengthy process of evacuation and cooldown after each adjustment. An alternative is to use a thicker window with enough resistance to bowing that its effects on focus are negligible.

An application where the pressure difference across the window *does* change is space-based sensors, where the pressure difference goes from approximately 100 kPa (14.7 psi) at sea level to essentially zero in orbit. If the *change* in OPD due to these changes is $>\lambda/4\,\mu m$ peak-to-valley, then the window is too big, too thin, or not stiff enough, and must be redesigned if there is no focus-adjust mechanism to accommodate the change in focus.

The peak-to-valley defocus OPD for a nearly simply supported window can be estimated from an equation derived by Barnes using the window's structural deflection [2]:

$$\text{OPD} \approx 0.01(n-1)\frac{D_w^6}{t^5}\left[\frac{\Delta P}{E}\right]^2 \qquad (6.1)$$

This equation shows that the OPD is worse (larger) for a larger window diameter D_w, a thinner window (via the thickness t), a flexible window (via the elastic modulus E), and a large pressure difference ΔP. For example, for a germanium window—a very stiff optical material with $E = 104\,\text{GPa}$ and a refractive index $n = 4$ for midwave infrared (MWIR) wavelengths—with $D_w = 100\,\text{mm}$ and $t = 3\,\text{mm}$, the defocus OPD $\approx 0.1\,\mu\text{m}$ for an evacuated Dewar. If such a Dewar is then launched into space, the OPD due to bowing is then zero. The *change* in OPD between ground and space is thus $0.1\,\mu\text{m}$, or 10 times $<\lambda/4$ for a wavelength $\lambda = 4\,\mu\text{m}$. This is much smaller than the Rayleigh defocus criterion, and a focus-adjust mechanism is fortunately not required for either ground or space-based operation for this example.

An OPD$>\lambda/4$ PV, however, does not automatically disqualify the window design. That is, because it can be removed at assembly, the OPD due to defocus can be treated differently than that due to spherical, astigmatism, and coma. If the defocus from this OPD is removed by correcting the focus at assembly, and there is no *change* in OPD—for example, if the pressure across the window does not change—then a large OPD may be perfectly acceptable, and the window design may not need to change. Thus, a maximum OPD of anywhere from $\lambda/4$ to 1λ may still be a good design criteria, although any larger OPD may result in imaging aberrations such as spherical (because of the change in surface curvature), coma (as the mount or loading will be asymmetric to some degree), or higher-order aberrations for thick windows or large incidence angles. Most importantly, the *change* in thickness Δt must be controlled to prevent an OPD$=(n-1)\Delta t$ and resulting WFE from being induced in the glass (Section 6.1.4).

6.1.2 Poisson's Ratio

Because the size of a plate is large in two dimensions—as distinct from a beam, which is only large in one—a new quantity known as Poisson's ratio must be introduced. Specifically, as a lens is pulled in one dimension—radially around its circumference, for example—its thickness shrinks in the other dimensions (Fig. 6.4). Similarly, if the lens is squeezed in one dimension, the thickness expands in the others. The amount of shrinkage or expansion is determined by a material parameter known as Poisson's ratio μ, defined as the ratio of strain in one dimension compared with the others ($\mu = \varepsilon_2/\varepsilon_1$ in Fig. 6.4). The shrinkage is a result of an atomic-coupling of the

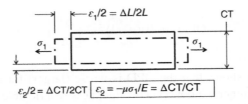

FIGURE 6.4 The strain $\varepsilon_2 = \Delta CT/CT$ across the center thickness (CT) is determined by the axial tensile strain $\varepsilon_1 = \sigma_1/E$ and Poisson's ratio μ.

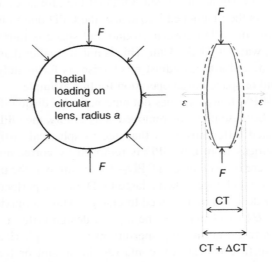

FIGURE 6.5 Radial loading around the circumference of a lens expands its thickness, as in the focus accommodation of the human eye.

strains and is approximately 20–30% ($\mu \approx 0.2$–0.3) of the tensile strain for structural materials.[1]

The shrinkage due to Poisson's ratio may change the surface shape and hence the WFE of a lens or window. This is shown in Figure 6.5, where a lens is radially loaded with a force F, and it would be useful to estimate the change in center thickness and resulting wavefront error. This is an extremely difficult problem and is usually solved via structural analysis of the type reviewed in Chapter 10. Using design handbooks such as *Roark's Formulas*, however, we can obtain a first-order estimate.

---

[1] Physically, all structural materials have some degree of atomic-force coupling between axes, such that pulling in tension also becomes a shear across atomic planes which results in Poisson's contraction.

Extracting the appropriate table for use from *Roark's Formulas* is not always straightforward, though we are fortunate in this case to find that a cylinder's longitudinal expansion due to a radial loading that is included in *Roark's* Table 13.5 [3]. We can view the "cylinder" in this case as a lens, whose radial loading results in an expansion of its length (Δl in Fig. 6.6), equivalent to a change in center thickness (ΔCT). If the lens radii are assumed to have negligible effect on the center-thickness expansion $\Delta l = \Delta$CT—as occurs for a window or weak lens, for example—then *Roark's Formulas* provides an approximation to the Poisson expansion.

Using Figure 6.6, the change in center thickness ΔCT $= 2\mu q$CT$/E = \mu F/\pi aE$ for $b=0$ and q is the pressure ($F/A \approx F/2\pi a$CT) due to the radial force F. For N-BK7 glass[2] with $\mu = 0.206$ and $E = 82$ GPa, we find that

$$\Delta \text{CT} = \frac{\mu F}{\pi aE} = \frac{0.206 \times 4450\,\text{N}}{\pi \times 0.025\,\text{m} \times 82\text{E}9\,\text{N}/\text{m}^2} = 0.14\,\mu\text{m}$$

for a 50-mm lens diameter ($a = 25$ mm) and $F = 4450$ N (1000 lbs). The resulting WFE $= (n-1)\Delta$CT $\approx 0.5 \times 0.14\,\mu$m $= 0.07\,\mu$m per 4450 N of radial load. For a lens used with a HeNe laser ($\lambda = 0.633\,\mu$m), this is $\sim\lambda/9$ PV of WFE.

Interpreting the results, it is not at all clear whether this WFE is simple defocus or something more complex. For example, if the surface is irregular—as it must be to some extent—then the irregularities will change in magnitude, as these slight variations in thickness will become larger as the lens is squeezed radially. The degree to which this occurs is more complex than this first-order analysis illustrates and may need to be analyzed using the structural-thermal-optical (STOP) methods reviewed in Chapter 10. Nonetheless, if the WFE from the first-order model is excessive, the radial load may need to be reduced, or the lens may need to be redesigned to something larger or stiffer.

6.1.3 Plate Bending

As we have seen, surface figure error (SFE) from mounting distortion is difficult to predict analytically. For this reason, the computational methods reviewed in Chapter 10 are used only when, as with space-based instruments, there is typically just one chance for success. For lower-budget commercial projects, an empirical approach—combined with experience and back-of-the-envelope calculations such as those developed in this book—are the norm.

---

[2] The SCHOTT website has a database of optical and mechanical properties for a large number of optical glasses such as N-BK7—see http://www.schott.com.

Case no., form of vessel	Case no., manner of loading	Formulas
1. Cylindrical disk or shell	1a. Uniform internal radial pressure q, longitudinal pressure zero or externally balanced; for a disk or a shell	$\sigma_1 = 0$ $\sigma_2 = \dfrac{qb^2(a^2+r^2)}{r^2(a^2-b^2)}$, $(\sigma_2)_{max} = q\dfrac{a^2+b^2}{a^2-b^2}$, at $r=b$ $\sigma_3 = \dfrac{-qb^2(a^2-r^2)}{r^2(a^2-b^2)}$, $(\sigma_3)_{max} = -q$, at $r=b$ $\tau_{max} = \dfrac{\sigma_2-\sigma_3}{2} = q\dfrac{a^2}{a^2-b^2}$, at $r=b$ $\Delta a = \dfrac{q}{E}\dfrac{2ab^2}{a^2-b^2}$, $\Delta b = \dfrac{qb}{E}\left(\dfrac{a^2+b^2}{a^2-b^2}+v\right)$, $\Delta l = \dfrac{-qvl}{E}\dfrac{2b^2}{a^2-b^2}$
	1b. Uniform internal pressure q, in all directions; ends capped; for a disk or a shell	$[\sigma_2,\sigma_3,(\sigma_2)_{max},(\sigma_3)_{max},$ and τ_{max} are the same as for case 1a] $\sigma_1 = \dfrac{qb^2}{a^2-b^2}$ $\Delta a = \dfrac{qa}{E}\dfrac{b^2(2-v)}{a^2-b^2}$ $\Delta b = \dfrac{qb}{E}\dfrac{a^2(1+v)+b^2(1-2v)}{a^2-b^2}$ $\Delta l = \dfrac{ql}{E}\dfrac{b^2(1+2v)}{a^2-b^2}$
	1c. Uniform external radial pressure q, longitudinal pressure zero or externally balanced; for a disk or a shell	$\sigma_1 = 0$ $\sigma_2 = \dfrac{-qa^2(b^2+r^2)}{r^2(a^2-b^2)}$, $(\sigma_2)_{max} = \dfrac{-q2a^2}{a^2-b^2}$, at $r=b$ $\sigma_3 = \dfrac{-qa^2(r^2-b^2)}{r^2(a^2-b^2)}$, $(\sigma_3)_{max} = -q$, at $r=a$ $\tau_{max} = \dfrac{(\sigma_2)_{max}}{2} = \dfrac{-q}{E}\dfrac{2a^2}{a^2-b^2}$, at $r=b$ $\Delta a = \dfrac{-qa}{E}\left(\dfrac{a^2+b^2}{a^2-b^2}-v\right)$, $\Delta b = \dfrac{-q}{E}\dfrac{2a^2}{a^2-b^2}$ $\boxed{\Delta l = \dfrac{qvl}{E}\dfrac{2a^2}{a^2-b^2}}$

Change in CT (Δl) as a lens is squeezed

FIGURE 6.6 Table 13.5 from *Roark's Formulas* provides an approximation to the Poisson expansion of center thickness (ΔCT) for a lens with a radial load. Adapted from Warren C. Young and Richard G. Budynas, *Roark's Formulas for Stress and Strain*, McGraw-Hill (2001).

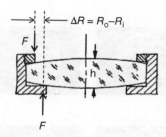

FIGURE 6.7 Offset axial loading creates a moment that bends an optical component. The radii R_o and R_i are the radii at which the force F is applied. Credit: Dan Vukobratovich [4].

In addition, there are a number of "best practices" and rules of thumb that can be used to minimize the distortion and WFE of optical components. For example, even seemingly simple decisions as to how a lens is mounted can distort its surfaces. For example, a lens retainer that is not aligned with the supporting surface behind it can distort the lens (Fig. 6.7).

The applied moment is $M = F \times \Delta R$. The maximum bending stress σ_{max} in the lens is [5]

$$\sigma_{max} = \frac{3\mu F}{2\pi h^2}\left[\frac{1}{2}\left(\frac{1}{\mu}-1\right)+\left(\frac{1}{\mu}+1\right)\ln\left(\frac{R_1}{R_o}\right)-\left(\frac{1}{\mu}-1\right)\left(\frac{R_o^2}{2R_1^2}\right)\right] \quad (6.2)$$

which shows that to reduce the stress and strain in the lens, the retaining ring should be co-aligned with the supporting surface such that $R_o = R_i$, from which Equation 6.2 gives $\sigma_{max} = 0$.

Another seemingly simple decision is the design of the retaining ring itself. This is shown in Figure 6.8, where the wavefront transmitted through a lens is shown with and without a retaining ring. It is seen that applying torque to the ring applies an asymmetric load to the lens, creating distortion and WFE fringe patterns similar to coma and astigmatism. By contrast, if the retaining ring is slotted (Fig. 6.8d), both the number and character of interference fringes is reduced, indicating a reduction in lens surface distortion. Depending on their depth, the slots thus give the retaining ring some amount of flexibility, reducing the deflection asymmetry of the lens surface.

6.1.4 Contact Stresses

Even if the retainer and its support are co-aligned and slotted, there will be additional stresses on the lens due to contact with the retainer. For example, a sharp knife edge used to retain the lens will impart very high stresses to the

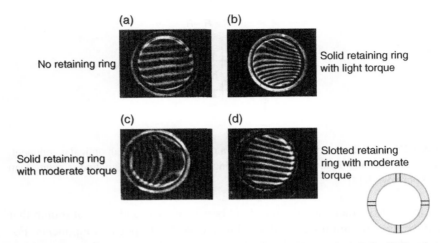

FIGURE 6.8 Slotting a retaining ring—as in the lower right—gives it flexibility that reduces the lens surface deformation and wavefront error. Photo credits: Mete Bayer, "Lens Barrel Optomechanical Design Principles", Optical Engineering, Vol. 20, No. 2 (1981).

FIGURE 6.9 The stress on a lens depends on the sharpness of the surface retaining it. Credit: Dan Vukobratovich, SPIE Press [5].

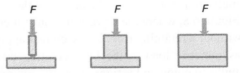

FIGURE 6.10 Which mirror will distort or break more easily?

lens because of the small area of the edge (Fig. 6.9). A large-area retainer, on the other hand, distributes the force over a larger surface, reducing the stress. Such stresses are called contact stresses, and reducing these stresses is critical in managing WFE and, in extreme cases, breakage.

The idea of contact stress is illustrated in Figure 6.10, where three concepts for retaining a mirror are compared: that which distributes the force F over a small area, over an intermediate area, and over the entire area of the mirror. From the definition of stress, we see that the smaller area results in a higher stress for a given load ($\sigma = F/A$). The mirror on the left in Figure 6.10 is thus more easily strained (and broken) than the other two, and such small-area contact is referred to as a contact stress.

Any stress that changes the thickness of a lens also changes the OPD and creates wavefront error. Contact stress can change the lens thickness in localized areas where the stress is applied. If these localized deformations extend into the clear aperture, then Figure 6.11 shows that the resulting PV SFE of Δt from *mounting strain* creates WFE. Fortunately, this may not invalidate the design, as it may be the root-mean-square (RMS) WFE over the entire area that determines optical performance (see Example 3.1).

Contact stress can thus be quantified by looking at how much the mating surfaces deform. In deforming elastically, the stress is reduced somewhat, in that the area supporting the load is larger than implied by the infinitesimal area of point or line contact singularities (Fig. 6.12). The deformation and

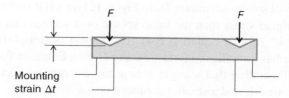

FIGURE 6.11 The load F compresses the optic over a small area, creating a change in thickness Δt and PV wavefront error WFE $= (n-1)\Delta t$ for a window or lens, or $2\Delta t$ for a mirror.

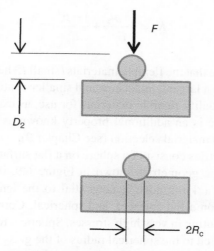

FIGURE 6.12 The curvature and elasticity of both surfaces determines the contact area and therefore the contact stress.

contact depends on the inverse stiffness, that is, the elastic flexibility $1/E$, of both surfaces, though the more flexible surface will of course dominate.

From Chapter 14 of *Roark's Formulas*, the contact-stress area $A = \pi R_c^2$, where Figure 6.12 shows that R_c is the contact-stress radius [3]:

$$R_c = 0.721 \sqrt[3]{FK_1K_2} \qquad (6.3)$$

The contact area—and compressive contact stress $\sigma_c = F/A$—thus depends on the force F, a geometry factor K_1, and a material factor K_2.

The geometry factor depends on the relative diameters (curvatures) of the two surfaces [3]

$$K_1 = \frac{D_2}{1 \pm D_2 / D_1} \qquad (6.4)$$

such that a small sphere (diameter D_2 in Fig. 6.12) on a flat surface of infinite radius has a higher stress than the same sphere on a surface that curves away from it (the "+" sign in Equation 6.4). The reason is straightforward: The curved surface has a smaller contact area. For $D_1 = \infty$, Equation 6.4 also shows that $K_1 = D_2$, illustrating that a larger sphere has a larger contact radius R_c, a larger contact area A, and a smaller contact stress σ_c.

The material factor K_2 depends on the elastic deformation of *both* parts in contact—a lens and its retainer, for example. Poisson's ratio also contributes, but it is the glass elasticity E_g and mount elasticity E_m that typically dominate [3]:

$$K_2 = \frac{1 - \mu_g^2}{E_g} + \frac{1 - \mu_m^2}{E_m} \qquad (6.5)$$

Physically, we see that the flexible materials (small E) have a larger material factor, and therefore a larger contact area and smaller contact stress. This does not automatically qualify flexible materials for use, as excess deflection creates WFE, and there is an additional property known as *hardness* that may also be important in material selection (see Chapter 9).

Not many geometries consist of a sphere on a flat surface. More typical for lens retainers are the geometries shown in Figure 6.9, illustrating retainers classified as corner ("knife edge"), tangential to the lens spherical surface (which is a sphere on a flat surface), and spherical. Corner retainers have a small load area, resulting in very high stresses. Spherical retainers are difficult to tolerance and match to the spherical radius of the glass surface. Because of their manufacturability and relatively low stress, the tangential retainer geometry shown in Figure 6.13 is common, as is the toroidal [6, 7].

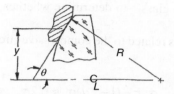

FIGURE 6.13 Geometry and nomenclature used to analyze the contact stress from a tangential retainer. Note that the surface radius R in this figure is analogous to $D_2/2$ in Figure 6.12. Credit: Dan Vukobratovich [4].

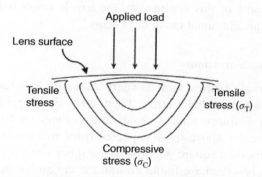

FIGURE 6.14 Compressive contact stresses result in tensile stresses in the area surrounding the applied load. Credit: Katie Schwertz, "Useful Estimations and Rules-of-Thumb for Optomechanics," University of Arizona MS Thesis (2010).

For a retainer force F holding a lens in place, the compressive stress σ_c in the lens for a tangential retaining ring is given by [6]

$$\sigma_c^2 = 0.637 \left[\frac{\dfrac{F}{4\pi y R}}{\dfrac{1-\mu_G^2}{E_G} + \dfrac{1-\mu_m^2}{E_m}} \right] \tag{6.6}$$

where Figure 6.13 shows that y is the distance from the lens centerline where the tangential contact is made, and R is the radius of the glass surface. Physically, we expect that the stress is lower for a larger surface radius R, flexible materials with small E, and a larger circumferential area $\sim 2\pi y$ to support the load—expectations confirmed by Equation 6.6.

Associated with compressive contact stresses is a surrounding tensile region to balance forces within the material (Fig. 6.14). As we will see in Section 6.2, it is these tensile stresses that must be compared with glass strength—approximately 7 MPa (1000 psi) for fine-ground glass or 14 MPa

(2000 psi) for polished glass—to determine whether a lens will survive the applied load.

The tensile stress σ_t is related to the compressive stress σ_c through Poisson's ratio μ [7]:

$$\sigma_t = \frac{1}{3}(1-2\mu)\sigma_c \approx 0.2\sigma_c \qquad (6.7)$$

Physically, atoms in compression must be surrounded by atoms that are being pulled in tension to accommodate the compression. Because Poisson's ratio counters some of this contraction, the tensile stress is lower than the compressive by an additional factor of $(1-2\mu)$.[3]

6.1.5 Stress Concentrations

We have seen that features such as sharp edges on lens retainers tend to elastically deform to reduce stress from their singular (infinite) value. However, other features such as holes and corners tend to amplify (or *concentrate*) stresses above those expected from area alone. For example, the sharp corners on a square window have higher stresses—and will thus break more easily—than are found around the circumference of a circular window (Fig. 6.15).

These stress concentrations are due to the redirection of stress patterns around a hole, crack, edge, or corner. As seen in Figure 6.16, the stress patterns are forced to concentrate around a crack that they cannot cross, thus increasing the stress locally. Qualitatively, a deeper crack or smaller-radius corner has higher stresses.

Sharp corners and edges may thus increase the stress in a window or lens beyond the material strength. Intentionally "breaking" the sharp corners of a square glass plate (Fig. 6.17)—also known as *chamfering* or *beveling*—reduces the stresses in the plate before the environment has a chance to do so

FIGURE 6.15 Which window will break more easily—the square, cornered, or round geometry?

---

[3] If there were no Poisson's ratio ($\mu = 0$), the tensile stress is lower than the compressive by a factor of 1/3. For a completely compressible material ($\mu = 0.5$), there are in principle no compensating tensile stresses.

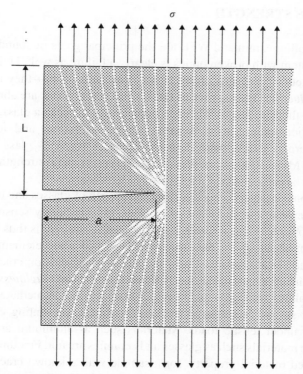

FIGURE 6.16 Stresses in a material are concentrated around the tip of a crack. Adapted from J. E. Shigley and C. R. Mischke, *Mechanical Engineering Design*, McGraw-Hill (1989).

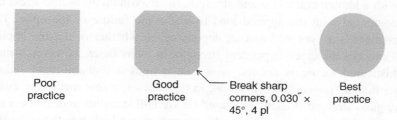

FIGURE 6.17 Stress concentrations at the sharp corners of a window require chamfering or beveling to reduce the stress at the corners.

catastrophically. As we saw in Chapter 3, chamfers are also used around the edge of a lens or mirror—also to reduce the stress concentrations. For more details on stress concentration factors, see *Roark's Formulas* [3] or *Peterson's Stress Concentration Factors* [8].

6.2   GLASS STRENGTH

While controlling strain and WFE are the principal goals of sound structural design, a system with broken optics is clearly a failure as well. Unlike optical materials, metals break gracefully under stress and strain—they are usually pliable and *ductile*. Glasses such as BK7, on the other hand, are abrupt in how they break—they are *brittle*. There are many ways to break a glass, crystal, or ceramic optic: static forces, dynamic shock, vibration, and temperature changes (thermal shock) being four examples. A "typical" glass strength is only about 7 MPa (1000 psi) of tensile stress; typical metal strength is approximately 50× larger, or ~0.35 GPa (50 ksi) of tensile stress.

The reason for the difference is that the strength of glasses and other brittle materials (ceramics, semiconductor crystals, etc.) is highly sensitive to surface flaws (or cracks); as a result, a simple tensile strength is thus both insufficient and misleading for the structural design of glass elements. What is needed instead is a measure of how the strength depends on crack size, and this is done using a material property known as *fracture toughness*.

A material that is tough—that is, resists cracks by not immediately breaking down under stress—has a large fracture toughness, preventing cracks from propagating under tensile loads. Materials such as aluminum and steel are tough, while materials such as glasses and ceramics are not. Fracture toughness is a measured material property, for a sample with a known crack size. The tensile stress at which the sample breaks (units of MPa) is combined with the measured crack size (units of $m^{1/2}$) to determine the fracture toughness (units of MPa-$m^{1/2}$). Fracture toughness can thus be used to determine a tensile "strength," whose value depends on crack size.

With a known crack size and strength, the maximum allowable stress can be compared with the applied load to assess the failure of the optic. This assessment is a "yes–no" metric, depending on whether or not the applied load exceeds the crack-dependent strength. In many cases, however, a more sophisticated metric is needed—one that assigns a probability of failure, rather than a "go–no go" evaluation. In these cases, a new method of evaluating glass strength is used, one based on Weibull statistics to describe a distribution of flaw sizes over an area. In this section, we look at both the fracture toughness and statistical approaches to glass strength.

6.2.1   Fracture Toughness

Using fracture toughness to estimate the strength of brittle optical materials is based on the idea that deeper cracks have a bigger stress concentration and will therefore break at a lower stress (Fig. 6.18).

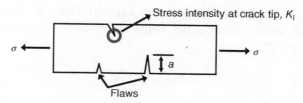

FIGURE 6.18 The surface flaws in brittle materials reduce the stress σ that can be applied before catastrophic failure occurs. Credit: Keith B. Doyle, Victor L. Genberg and Gregory J. Michels, *Integrated Optomechanical Analysis* (2nd ed.), SPIE Press (2012).

TABLE 6.1 Fracture Toughness K_{IC} of Commonly Used Brittle Materials[a]

Material	K_{IC} (MPa-m$^{1/2}$)	K_{IC} (psi-in.$^{1/2}$)
Zinc selenide (ZnSe)	0.5	455
Germanium (Ge)	0.7	637
Fused silica 7940 (SiO$_2$)	0.74	675
ULE	0.75	683
Silicon (Si)	0.9	820
Zerodur	0.9	820
Zinc sulfide (ZnS)	1.0	910
N-BK7	1.1	1000
Sapphire (Al$_2$O$_3$)	2.0	1820
Diamond	3.4	3094
Silicon carbide (SiC)	4.0	3640

[a] Data are from Yoder [9] and SCHOTT [10].

Quantitatively, brittle fracture for glasses, crystals, and optical ceramics occurs when the stress intensity factor K_I (units of MPa-m$^{1/2}$ or psi-in.$^{1/2}$) exceeds the fracture toughness K_{IC}—a measured material parameter with the same units as K_I:

$$K_I > K_{IC} \qquad (6.8)$$

A larger value of K_{IC} requires more energy to propagate cracks in a material, and for the same crack size can withstand a higher stress than a material with lower fracture toughness. Measured values are given in Table 6.1; typical numbers for glasses are $K_{IC} \approx 1$ MPa-m$^{1/2} \approx 900$ psi-in.$^{1/2}$. Fracture toughness may vary by a factor of 2 or more from those listed in the table.

As shown in Equation 6.9, the stress intensity factor K_I depends on the applied tensile stress σ, crack depth a (also known as flaw depth), and a crack geometry factor[4] Y:

$$K_I = \sigma Y \sqrt{a} \qquad (6.9)$$

Due to the stress concentration (Fig. 6.16), deeper cracks (larger a) reduce the value of the applied tensile stress σ needed to maintain $K_I < K_{IC}$. Crack depth and maximum applied stress—that is, the *fracture strength*—thus trade against each other; a simple measurement of fracture strength is thus insufficient for specifying glass properties, without also specifying crack size.

The crack geometry factor Y depends on width-to-depth ratio of crack, with values on the order of 1 to 2 (for a width-to-depth ratio <5 to >10), and $Y \approx 1.25$ is typically used in first-order design calculations.

Table 6.1 shows that commonly used brittle materials include fused silica, germanium, and zinc selenide (ZnSe). High fracture-toughness materials such as sapphire, silicon carbide (SiC), and diamond require more energy to propagate cracks and thus break at a higher stress for a given crack size. For comparison, the fracture toughness of metals is much higher than those of even tough ceramics such as SiC, with aluminum's $K_{IC} \approx 40\,MPa\text{-}m^{1/2}$, and that of a good alloy steel is $\approx 150\,MPa\text{-}m^{1/2}$.

Silicon carbide (SiC) is a structural ceramic becoming more common in optical systems. While silicon carbide has a relatively large fracture toughness—higher than that of diamond—and is very stiff compared with its weight, its brittleness has slowed down its use. Recent space-based instruments are using SiC with a better understanding of its fracture toughness and its manufacturing variations—see the European Space Agency website (www.esa.int) for details.

Because fracture strength depends on crack depth a, the allowable stress can be increased by reducing the size of the surface flaws. Specifically, the crack depth depends on the size of the abrasive particles used to grind and polish the material (Fig. 6.19); smoothly polished surfaces with smaller a can thus withstand higher stresses before breaking. Common cerium oxide polishing compounds have particle sizes on the order of $1\,\mu m$ (or $0.04 \times 10^{-3}\,in. \equiv 0.04\,mil$). However, glass strengths of up to 1000× stronger than typical—as high as $6.9\,GPa$ ($10^6\,psi$)—have been obtained not by polishing but by acid etching of surface flaws; glass manufactured for smartphone displays sometimes use this approach to increasing strength. In addition, the strength of Corning's smartphone Gorilla Glass and common auto windshields is increased not by controlling the crack size but by manufacturing the glass

---

[4] In the literature, there are other values of fracture toughness based on the load geometry. For example, K_{IIC} and K_{IIIC} are used for shear loads—see Ref. [1].

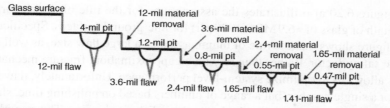

FIGURE 6.19 The flaw depth $a \approx 3 \times$ the size of the visible pit from the size of grinding and polishing particles. Credit: Keith B. Doyle, Victor L. Genberg and Gregory J. Michels, *Integrated Optomechanical Analysis* (2nd ed.), SPIE Press (2012).

with an inherent compressive stress in the surface; such *tempered* glass thus increases the allowable stress in comparison with the tensile stresses to which the glass is much more sensitive to breakage.

Example 6.1 This example illustrates how crack depth and fracture strength trade against each other. Combining Equations 6.8 and 6.9, the tensile fracture strength $\sigma_f = K_{IC}/Ya^{1/2}$. In addition, because brittle materials fracture catastrophically into multiple shards, a generous safety factor (SF) is used in glass strength calculations, thus reducing the fracture strength used for design by the SF. The fracture strength for N-BK7 (with a toughness $K_{IC} = 1.1\,\text{MPa-m}^{1/2}$) is shown in Figure 6.20 for a range of flaw sizes and a safety factor SF=4. The results place an upper limit on the applied load for a given flaw size. Alternatively, the results put an upper limit on flaw size for a given stress, again illustrating the design trade between strength and flaw size.

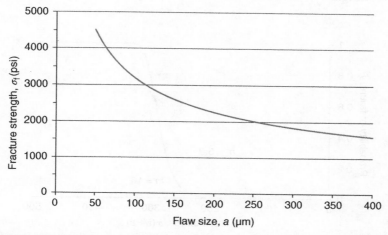

FIGURE 6.20 Fracture strength σ_f decreases as surface flaw size a increases. Assumptions for the plot are a safety factor SF=4 and geometry factor $Y=1.25$.

Figure 6.20 also illustrates the assumptions for the rule of thumb for the strength of glass of ≈6.9 MPa (1000 psi) for fine-ground surfaces. Specifically, the figure shows that the rule of thumb assumes a large flaw size, as well as a large safety factor. Rather than rely on an approximation, fracture mechanics thus allows us to optimize system-level performance. Unfortunately, flaw size is *not* a single number, but a range of numbers based on polishing time, slurry pH, and the particle-size distribution in the polishing abrasive. This complication is addressed in the next section.

6.2.2 Weibull Statistics

Because flaw size from the fabrication process is somewhat random, the estimate of fracture strength from a single crack depth is not complete. In addition, the number of flaws depends on the surface area—a complexity not captured by fracture-toughness calculations. As a result, it is necessary to take into account the statistical distribution of flaw sizes and number of flaws that may occur in the polishing and fabrication of optical components.

Weibull statistics are used to address these shortcomings by calculating a probability of failure P_f, rather than the digital "yes-or-no" failure criterion implied by fracture mechanics [1]:

$$P_f = 1 - \exp\left[-\left(\frac{\sigma}{\sigma_o}\right)^m\right] \qquad (6.10)$$

This equation is plotted in Figure 6.21 for an applied tensile stress σ, a characteristic strength σ_o (where 63.2% of tested samples will fail, i.e., there is a 63.2%

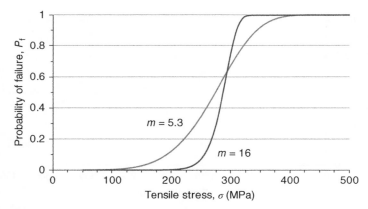

FIGURE 6.21 Comparison of the probability of failure for two materials with the same characteristic strength $\sigma_o = 293.8$ MPa but a different Weibull modulus m.

TABLE 6.2 Probability of Failure at $\sigma = 6.9\,$MPa (1000 Psi) Using Equation 6.10 for Common Optical Materials[a]

Material	Characteristic strength, σ_o (MPa)	Weibull modulus, m	Probability of failure, P_f
N-BK7	70.6	30.4	0.0
F2	57.1	25.0	0.0
SF6	57.3	21.9	0.0
Zerodur	293.8	5.3	2.5×10^{-9}
Silicon	346.5	4.5	2.2×10^{-8}
Sapphire	485	4.0	4.1×10^{-8}
ZnSe	54.9	6.0	3.9×10^{-6}
Germanium	119.8	3.4	6.1×10^{-5}
ULE	40.4	4.5	3.8×10^{-4}
Calcium fluoride	5.0	3.0	0.93

[a] Material data for σ_o and m from Ref. [6]; surface area and mean surface finish varies for each material.

probability of failure), and a Weibull modulus m (where σ_o and m both depend on the scratch depth and surface area). The larger Weibull modulus shows a greater sensitivity to changes in tensile stress so if there are large uncertainties in the stress calculation, it is best to use a material with a small m. The smaller modulus is much more tolerant to changes in the tensile stress near the characteristic strength, avoiding a risky situation where small analysis uncertainties result in large changes in failure probability.

Table 6.2 shows the probability of failure for commonly used brittle materials at a stress of 6.9 MPa (1000 psi). The value of 6.9 MPa was selected to give us a reference to the rule-of-thumb regarding glass strength for fine-ground surfaces [11]. The table shows that some materials will clearly survive, while others (calcium fluoride) will surely fail. The strength rule-of-thumb of 6.9 MPa is thus exactly that—an approximation that should be replaced with more exact analysis using Weibull statistics whenever possible.

Both characteristic strength σ_o and Weibull modulus m are measured parameters whose value depends on the surface finish. Zerodur, for example, has values of $\sigma_o = 293.8\,$MPa and $m = 5.3$ for a surface with a conventional optical polish [10], while the strength drops to $\sigma_o = 108\,$MPa and the modulus increases to $m = 16$ for a larger-particle SiC 600 polish (using silicon carbide polishing grit with a mean particle size of 9 µm and flaw depth $a \approx 30$ µm—see Fig. 6.19). While it is not clear how m depends on finish, characteristic strength scales with depth a of the surface cracks as follows [10]:

$$\frac{\sigma_o}{\sigma_{o,\text{design}}} = \left[\frac{a_{\text{design}}}{a}\right]^{1/2} \qquad (6.11)$$

Deeper scratches than those present when measuring σ_0 have a higher stress concentration, resulting in a smaller characteristic stress $\sigma_{o,design}$. Thus, scratches with $a_{design} > a$ cause failure at a design tensile stress $\sigma_{o,design} < \sigma_0$. Both polished and unpolished areas must be checked for strength, as unpolished areas—fine- or medium-grind edges, for example—may have deeper scratches and thus lower strength.

As a larger surface area also has a higher probability of containing critical crack flaws, characteristic strength also scales with surface area [10]:

$$\frac{\sigma_0}{\sigma_{o,design}} = \left[\frac{A_{design}}{A} \right]^{1/m} \tag{6.12}$$

Larger areas than used when measuring σ_0 have more scratches, resulting in a smaller characteristic stress $\sigma_{o,design}$. Larger optics thus fail at a lower tensile stress. For example, SCHOTT Glass uses an area $A = 113\,mm^2$ for their lab measurements for σ_0; a larger area than this value in your design will result a lower characteristic strength $\sigma_{o,design}$.

Example 6.2   This example illustrates the effects of surface area on the fracture strength of an optic. Specifically, a large zinc selenide (ZnSe) window is 500 mm in diameter and must be designed to have a probability of failure of 10^{-4}. This requires keeping the applied stress low enough for the given area and scratch depth. How low does this stress need to be?

Equation 6.10 can be solved for the fracture stress σ, giving $\sigma = \sigma_0 [-\ln(1 - P_f)]^{1/m}$. The surface area used to measure the characteristic strength σ_0 for ZnSe is $A = 579\,mm^2$. The window, however, has a surface area $A_{design} = 0.25\pi(500\,mm)^2 = 1.96E5\,mm^2$, which is larger than data measured with $A = 579\,mm^2$, so the design strength $\sigma_{o,design}$ decreases. How much it decreases is given by Equation 6.12, such that $[A_{design}/A]^{1/m} = [1.96E5\,mm^2/579\,mm^2]^{1/15.7} = 1.45$; as a result, $\sigma_{o,design} = \sigma_0/1.45$. This assumes the modulus m does not also scale with surface area. Also note that the surface area of only one side of the window is used in Equation 6.12 to scale the characteristic strength—see Problem 6.8.

Summarizing: glass strength can in general be increased by either reducing the surface area [optical clear aperture (CA) and edge thickness] or controlling the crack depth. Methods for controlling strength include the following:

- Fabrication using fine polish (optical CA) or fine grind (edges);
- Use liquids and a soft cloth to clean optics; avoid rubbing that creates scratches;

TABLE 6.3 Acid Etching Flaws from a Surface of N-BK7 Increases σ_0 and Decreases M^a

Material	Surface condition	Characteristic strength, σ_0 (MPa)	Weibull modulus, m	Maximum particle size (um)
N-BK7	SiC 600	70.6	30.4	19
N-BK7	D64	50.3	13.3	63
N-BK7	D64, acid etch	234.7	4.1	63

^{a}Credit: "Design strength of optical glass and Zerodur", Technical Note TIE-33, SCHOTT Glass (2004).

- Avoid sharp metal surfaces on glass, as they can produce scratches and increase the contact stress on the glass—see Section 6.1.4;
- Use chamfers or bevels to avoid stress concentrators such as sharp corners and edges of the glass itself—see Section 6.1.5;
- Surface etching with acid—see Table 6.3 for effects on σ_0 and m for N-BK7;
- Chemical surface treatment to leave the surface in compression—for example, Gorilla Glass;
- Use hardened protective surface coatings—for example, diamond.

An additional complication in evaluating glass strength is that the crack size of stressed components increases in a corrosive environment, which apparently includes humid air due to its acidity. So the Weibull probability of failure is not a constant number, but increases over time when used in a humid environment [1]. All glasses break eventually under tension; the question is: how long it will take, given the environment?

In general, glass selection is initially based on an optical engineer's estimate of key optical properties such as transmission [12]. Following that assessment, it is the optomechanical engineer's responsibility to evaluate the material's mechanical properties, such as the ability to withstand both breakage and excess surface deformation. It is likely that structural deformation will be the driver of the design, with stiffness (E) and specific stiffness (E/ρ)—see Table 6.4—playing dominant roles. The specific stiffness is particularly important for windows, which are not typically loaded with forces other than their own weight. A vibration environment where the window is exposed to random vibration forces, for example, can cause the window to dynamically oscillate in a manner similar to a vibrating membrane or drumhead, affecting the performance of the subsequent optics. Minimizing these vibration effects requires a window with a large E/ρ, a topic we consider in detail in Chapter 7.

TABLE 6.4 Mechanical Properties of Typical Brittle Materials Used in Optical Systems, including Mirror Substrates and Refractives. Note that a Value for Fracture Strength is not Listed[a]

Material	Density, ρ (kg/m³)	Elastic modulus, E (GPa)	Specific stiffness, E/ρ (m²/s²)
Mirror substrates			
Borosilicate glass	2230	63	28.3×10^6
Silicon carbide (CVD)	3210	466	145×10^6
ULE	2210	68	30.8×10^6
Zerodur	2530	91	36×10^6
Refractives			
Fused silica	2200	72	32.7×10^6
Germanium	5330	104	19.5×10^6
N-BK7	2510	82	32.7×10^6
Sapphire	3970	335	84.4×10^6
Silicon	2330	131	56.2×10^6
Zinc selenide (ZnSe)	5270	67.2	12.8×10^6
Zinc sulfide (ZnS)	4080	74.5	18.3×10^6

[a] Data are from Ref. [6] and Crystran's *Handbook of Optical Materials* (www.crystran.co.uk).

PROBLEMS

6.1 In Section 6.1.1, it was stated that a pressure difference across a flat window bows the window into a weak negative lens (i.e., a lens with a negative focal length, and little refractive power). Show that this is the case using the simple lens formula for the focal length of a thin lens (see Chapter 2).

6.2 In Section 6.1.4, it was stated that a small sphere on a flat surface of infinite radius has a higher stress than the same sphere on a surface that curves away from it. Is the stress higher or lower when the mating surface curves toward the small sphere?

6.3 Is a tangential retainer possible for the concave side of a lens? If not, how is a lens with a concave surface retained?

6.4 Are brittle materials the same as those with low fracture toughness (i.e., a strength that is very sensitive to surface cracks)? Can you imagine a material that is tough, yet breaks in a brittle manner?

6.5 For a material with a fracture toughness $K_{IC} = 1.0$ MPa-m$^{1/2}$ and an applied stress of 6.9 MPa (1000 psi), what crack size is allowed before failure occurs? How does this crack size compare with those obtained in polishing the surface using a cerium abrasive?

6.6 For the plot shown in Figure 6.20, what flaw size is required to obtain a fracture strength σ_f of 10^6 psi? Does the flaw size obtained sound reasonable? How can such a flaw size be obtained?

6.7 From the perspective of reducing risk for life-critical window designs such as the space station or bathyspheres, which is preferred: a large or small Weibull modulus m?

6.8 In Example 6.2, the surface area of only one side of the window was used to scale the parameter σ_o. Why do we not use the area of both sides of the window? *Hint*: see Figure 5.12.

6.9 How would you protect glass from the impact of sand and dirt?

6.10 A trend seen in Table 6.3 is that the change from SiC 600 abrasive to D64 decreases the Weibull characteristic strength σ_o. Is this reasonable?

REFERENCES

1. K. B. Doyle, V. L. Genberg, and G. J. Michels, *Integrated Optomechanical Analysis* (2nd Edition), Bellingham: SPIE Press (2012).

2. W. P. Barnes, "Optical windows," in *Optomechanical Design, SPIE Critical Reviews*, Vol. 43, Bellingham: SPIE Press (1992).

3. W. C. Young and R. G. Budynas, *Roark's Formulas for Stress and Strain*, New York: McGraw-Hill (2001).

4. D. Vukobratovich, *Introduction to Optomechanical Design*, SPIE Short Course SC014 (www.spie.org) (2009).

5. D. Vukobratovich, "Optomechanical system design," in M. C. Dudzik (Ed.), *The Infrared and Electro-Optical Systems Handbook*, Vol. 4, Bellingham: SPIE Press (1993), Chap. 3.

6. K. Schwertz and J. H. Burge, *Field Guide to Optomechanical Design and Analysis*, Bellingham: SPIE Press (2012).

7. K. J. Kasunic, J. Burge, and P. Yoder, Jr., *Mounting of Optical Components*, SPIE Short Course SC1019 (www.spie.org) (2013).

8. W. D. Pilkey and D. F. Pilkey, *Peterson's Stress Concentration Factors*, Hoboken: John Wiley & Sons, Inc. (2008).

9. P. R. Yoder, Jr., *Opto-Mechanical Systems Design* (3rd Edition), Boca Raton: CRC Press (www.crcpress.com) (2005).

10. SCHOTT Glass, "Design strength of optical glass and Zerodur," Technical Note TIE-33 (www.schott.com) (2004).

11. R. Williamson, *Field Guide to Optical Fabrication*, Bellingham: SPIE Press (2011). On p. 26 of this reference, it is stated that the typical abrasive stages are 30 μm or 20 μm (course grind), 15 μm or 12 μm (medium grind), and 9 μm or 5 μm (fine grind).

12. K. J. Kasunic, *Optical Systems Engineering*, New York: McGraw-Hill (2011).

7

STRUCTURAL DESIGN—VIBRATIONS

In an unfortunate display of nature's ability to eventually destroy anything that man can create, a steel bridge over the Tacoma Narrows was knocked down on a windy day in 1940 (Fig. 7.1). What's unusual about this event is not that wind can destroy a steel structure, but that the wind was relatively slow, on the order of only 40 miles/h (64 km/h). Despite the low speed, the wind set up twisting motions of the structure, and the frequency of these motions closely matched the structure's inherent ("natural") torsional frequency. Such a situation can result in self-induced structural motions that are larger than those predicted by static forces alone—large enough to bring an otherwise well-engineered structure tumbling down [1, 2].

In optical systems—and modern bridge designs as well—the effects of vibrations can be controlled to prevent destruction. From the perspective of imagers, radiometers, and similar optical instruments, the vibration-induced motions will more likely create dynamic misalignments between optical components, as well as changes in the line-of-sight (LOS) pointing direction (Fig. 7.2). Similarly, laser "technical" noises are power and wavelength fluctuations due to vibration-induced, sub-wavelength changes in laser cavity length or alignment.

The requirement on reducing the effects of vibrations thus depends on the application (Table 7.1). High-precision optical instruments such as

Optomechanical Systems Engineering, First Edition. Keith J. Kasunic.
© 2015 John Wiley & Sons, Inc. Published 2015 by John Wiley & Sons, Inc.

FIGURE 7.1 The Tacoma Narrows Bridge as it was destroyed by a relatively slow wind. Credit: K. Y. Billah and R. H. Scanlan, "Resonance, Tacoma Narrows bridge failure, and undergraduate physics textbooks," American Journal of Physics, Vol. 59, No. 2, pp. 118–124 (1991).

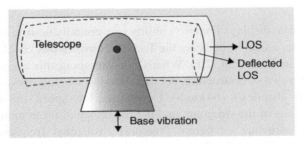

FIGURE 7.2 Vibrations of a telescope can create changes in the direction (or line-of-sight, LOS) along which the telescope optics are pointing. Adapted from J. M. Hilkert, "Inertially Stabilized Platform Technology," IEEE Control Systems Magazine, Feb. 2008.

laboratory-size interferometers, for example, are highly sensitive to changes in cavity length—a result of the wavelength-level precision needed for interference effects. The Laser Interferometer Gravitational-Wave Observatory (LIGO) built for the detection of gravity waves, where the effects of optical misalignment are propagated over a 4000-m optical path length, has an ever tighter requirement—one of the smallest of any instrument built to date.

Sources of vibration include structural motion of buildings (due to wind, traffic, people, etc.), nearby machinery and motors, air motion, platform motion (such as helicopters, satellites, cars, etc.), or anything that moves, *even if that motion is not itself vibratory*. Figure 7.3 shows a typical laboratory environment, which can be among the most benign, but still impose difficult requirements on the optomechanical system.

TABLE 7.1 The Resolution or Detail Size of Common Applications Determines the Requirements on the Vibration-control System[a]

Details Size (μm)	Description of Use
N/A	Distinctly discernible vibration. Appropriate to workshops and nonsensitive areas
N/A	Discernible vibration. Appropriate to offices and nonsensitive areas
75	Barely discernible vibration. Probably adequate for computer equipment, probe test equipment and low-power (to 20×) microscopes
25	Vibration not discernible. Suitable in most instances for microscopes to 100×
8	Adequate for most optical microscopes to 400×, microbalances, optical balances, proximity and projection aligners
3	Appropriate for optical microscopes to 1000× inspection and lithography equipment (including steppers) to 3 micron line widths
1	A good standard for lithography and inspection equipment to 1 µm detail size
0.3	Suitable for the most demanding equipment, including electron microscopes (TEMs and SEMs) and E-beam systems.
0.1	A difficult criterion to achieve in most instances. Assumed to be adequate for long-path laser-based interferometers and other systems requiring extraordinary dynamic stability

[a]Credit: CVI Laser, LLC.

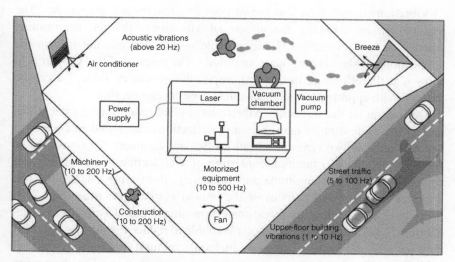

FIGURE 7.3 Common sources of vibration include electrical and mechanical machinery (motors, vacuum pumps, etc.), air flow, moving people and cars, and so on. Credit: CVI Laser, LLC.

TABLE 7.2 Common Industrial Sources of Vibration and their Associated Frequencies and Amplitudes[a]

Common Vibrational Sources

Source	Frequency (Hz)	Amplitude (in.)
Air Compressors	4–20	10^{-2}
Handling Equipment	5–40	10^{-3}
Pumps	5–25	10^{-3}
Building Services	7–40	10^{-4}
Foot Traffic	0.55–6	10^{-5}
Acoustics	100–10,000	10^{-2} to 10^{-4}
Air Currents	Labs can vary depending on class	Not applicable
Punch Presses	Up to 20	10^{-2} to 10^{-5}
Transformers	50–400	10^{-4} to 10^{-5}
Elevators	Up to 40	10^{-3} to 10^{-5}
Building Motion	46/height in meters, horizontal	10^{-1}
Building Pressure Waves	1–5	10^{-5}
Railroads	5–20	±0.15g
Highway Traffic	5–100	±0.001g

[a] Credit: CVI Laser, LLC.

Vibration sources have a range of both frequencies and amplitudes. The requirement on reducing the effects of vibrations—that is, determining whether or not a vibration is "discernable"—thus depends not just on the application but also on the source amplitude. The purpose of this chapter is to show that the effects of the typical amplitudes shown in Table 7.2 can be reduced with appropriate optomechanical systems engineering.

Depending on the frequency, vibrations applied to an optical system can deflect the system more than the static deflections calculated in Chapter 5 and 6. There are two types of vibrations that are common: (1) sinusoidal or "sine sweep" and (2) random, consisting of a collection of many different sinusoids of varying amplitude and frequency. Both types change the tilt, decenter, despace, and defocus of the optical system over time (Fig. 7.4). Sections 7.1–7.3 allow us to estimate how much the alignments change. Sections 7.4–7.6 review the methods available to reduce the misalignments if they are excessive; these methods include structural stiffening, damping, balancing, vibration isolation, and motion compensation. The discussion starts in Section 7.1 with the simplest case: the sinusoidal vibration of a mass-spring-damper system.

7.1 SINUSOIDAL VIBRATIONS

A simple optomechanical systems consists of a single mass (such as a beam), a spring (such as the beam's inherent stiffness), and an energy-damping element (such as friction). Mass–spring–damper systems have a natural frequency of vibration (Fig. 7.5); when the applied frequency matches this natural frequency—a situation known as *resonance*—the mass can vibrate with a bigger amplitude than its static displacement. As a result, the tilt, decenter, etc. can be significantly larger than expected. To understand why, we first look at natural frequencies and how the displacement of the simple mass–spring–damper system depends on the sinusoidal frequency applied to it.

7.1.1 Free Vibrations

The simplest mass–spring–damper system has little damping to dissipate energy (Fig. 7.6) and can freely vibrate (or oscillate) at a natural frequency ω_o which depends on the spring stiffness k and the mass m

$$\omega_o = \sqrt{\frac{k}{m}} \quad [\text{rad/s}] \quad (7.1)$$

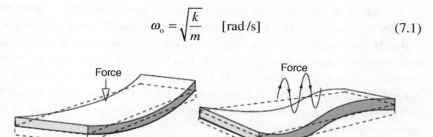

FIGURE 7.4 Vibration forces can distort the optomechanical structure holding the optics, creating tilt, decenter, despace, and defocus in ways that are not obvious from the static deflection shown on the left. Credit: CVI Laser, LLC.

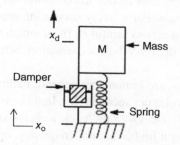

FIGURE 7.5 A simple optomechanical system can be broken down into mass, spring, and damping components. Permission to use granted by Newport Corporation; all rights reserved.

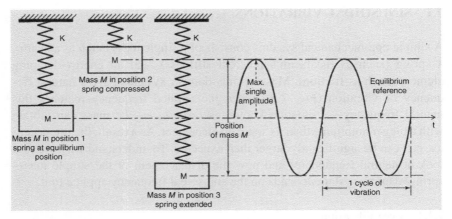

FIGURE 7.6 The natural frequency $\omega_0 = 2\pi/T$ of a mass-spring system depends on the sinusoidal period T, which is the time required for one cycle of vibration. Credit: Barry Controls.

A result of an application of Newton's second law, Equation 7.1 shows that a heavy mass vibrates at a lower frequency, while a stiff spring vibrates at a higher frequency. Physically, a larger mass has more inertia to be accelerated as it changes direction during vibration; this requires a stiffer spring—which provides a larger spring force $F = kx$ for a mass displacement x—to keep the natural frequency the same. Equivalently, a heavier mass takes longer to vibrate—and thus has a lower natural frequency—if the spring stiffness is not increased to compensate.

The source of Equation 7.1 is Newton's second law, where the acceleration of the mass depends on the sum of the forces acting on it (Fig. 7.7). For a spring force F which depends linearly on how much it is stretched or compressed—that is, $F = -kx$ for a spring stiffness k and displacement x—Newton's force balancing tells us that the equation $ma + kx = 0$ determines the acceleration a of the mass. For a freely oscillating mass, the time-dependent sinusoidal displacement $x(t) = x_0 \sin(\omega_0 t)$, from which we obtain $m\omega_0^2 - k = 0$ using $a = d^2x(t)/dt^2$ and the force-balance equation. Simple algebra then gives us Equation 7.1.

While the units of ω_0 are radians per second, natural frequencies are also expressed in units of cycles per second (Hz). In this case, the symbol f_0 is used instead, where $\omega_0 = 2\pi f_0$ (rad/s = 2π rad/cycle × cycles/s).

In addition to having a higher natural frequency, optomechanical systems that are lighter and stiffer also vibrate with smaller amplitude. Physically, structures with a higher natural frequency do not have the time to oscillate

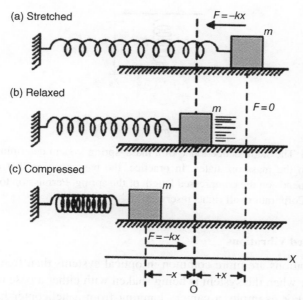

FIGURE 7.7 The derivation of Equation 7.1 relies on the application of Newton's second law for a mass m, spring force $F=-kx$, and a displacement $\pm x$. Permission to use granted by Newport Corporation; all rights reserved.

with large amplitudes. That is, the large f_o structure must quickly return to center, which it cannot do if its amplitude is large.

For a vibration in the vertical direction against gravity, the center (or offset) point of the vibration is the static deflection x_s (Fig. 7.8); all vibrations then occur around this static offset. Using Equation 7.1 and again applying Newton's second law, the static deflection also varies inversely with the natural frequency

$$x_s = \frac{F}{k} = \frac{mg}{k} = \frac{g}{\omega_o^2} \qquad (7.2)$$

This equation connects a static property we saw in Chapter 5 with the dynamic property of natural frequency ω_o. It also illustrates the concept of *gravity release*, where an optical system aligned on the surface of the Earth—where $g \approx 9.86\,\text{m/s}^2$—will have misalignments once launched into a space environment where $g \approx 0$. The equation also shows that a lighter (smaller m), stiffer (higher k) system with higher ω_o is preferred to keep the static deflections such as optical tilt and decenter low. What is not yet clear, however, is how much the *dynamic* deflections are affected by the natural frequency. For this, we need to introduce the concept of forced vibrations.

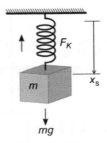

FIGURE 7.8 The static deflection x_s of a mass–spring system determines the offset around which the mass oscillates. In practice, the base position that defines zero deflection depends on the compressed length of the spring. Permission to use granted by Newport Corporation; all rights reserved.

7.1.2 Forced Vibrations

Forced vibrations are more common in optical systems than free vibrations. These occur when the system is being shaken with either a base or mass displacement. For example, a camera hanging from a helicopter has its mass shaken by the wind, while the camera also has its base vibrated by the same helicopter platform on which it is mounted—see Figure 7.5, where the "base" is the ground plane on which the spring and damper are mounted. For a sinusoidal base displacement $x_o(t)=x_0 \sin(\omega t)=x_o \sin(2\pi f t)$, the amplitude x_d of the mass depends on the comparison of the applied (or "driving") frequency f with the system's natural frequency f_o.

For $\omega/\omega_0 = f/f_0 \ll 1$—that is, a low-frequency situation—the mass closely follows the base displacement at the same frequency. For $f/f_0 \approx 1$—that is, a "resonance" situation, where the applied frequency resonates with the natural frequency of the structure—the mass displacement can be larger than the base displacement (Fig. 7.9).

The reason for the larger displacement is that at a natural frequency, the applied force F is in the same direction as the velocity $\mathbf{v}$ [3]. The resulting work applied to the mass ($\mathbf{F} \cdot \mathbf{v}$) efficiently increases its amplitude beyond the low-frequency displacement in a way that can cause bridges to fail, or optical components to misalign.

The amplitude when the applied frequency $f \approx f_0$ cannot increase without bound; instead, the limitation on resonance amplitude is the dissipation of energy by a concept known as "damping." The effects of this quantity are shown in Figure 7.9 by the varying values of the parameter ξ; Section 7.1.3 has more details.

Figure 7.10 shows a free-body diagram for the situation of a mass displacement such as a wind-driven telescope. Shown are the spring force ky, the

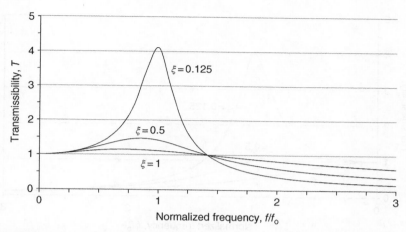

FIGURE 7.9 When the applied frequency f is near the natural frequency f_o of a mass–spring–damper system, the mass can move with an amplitude x_d that is larger than the base input amplitude x_o, giving a transmissibility $T \equiv x_d/x_o > 1$. (See Section 7.1.3 for the definition of ξ.)

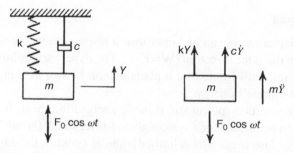

FIGURE 7.10 Schematic and free-body force diagram for the situation of a mass displacement, with the applied force F_o acting directly on the mass m with frequency ω.

damping force $c \times dy/dt = cv$, and the inertial force $m \times d^2y/dt^2 = ma$. Physically, the damping and inertial forces are small at low frequencies, the spring force ky balances the applied force F_o, and the force and displacement are in phase ($\varphi = 0$). At high frequencies, the applied force balances (mostly) the inertial force, so the force and displacement are out of phase ($\varphi \approx \pi$). At a resonant frequency, the applied force and *displacement* are in quadrature ($\varphi \approx \pi/2$). That is, the applied force and the velocity are in phase at resonance, just as with the case of a base displacement. The associated resonance curve for mass displacement is shown in Figure 7.11. In this case, the near-resonant deflection is similar to that of a base displacement, but the higher frequencies have somewhat different properties.

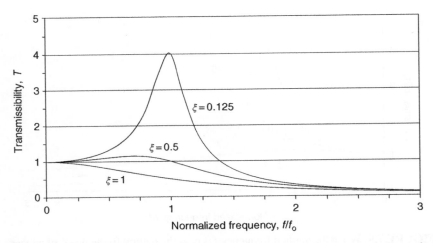

FIGURE 7.11 The mass displacement of a mass–spring–damper system can also force the mass to move with an amplitude x_d that is larger than the input amplitude x_o (i.e., $T > 1$).

7.1.3 Damping

The output displacement can be larger near a resonance, where the force and velocity are in the same direction ($W = \mathbf{F} \cdot \mathbf{v}$). The dynamic amplitude δ_d is thus able to build up, until limited by a phenomenon known as *damping,* which dissipates the motion's energy.

Damping converts displacement (kinetic energy) into heat. It is measured as a decrease in amplitude for a system given a brief shock ("rung") or impulse (Fig. 7.12). For low-stress (strain limited) optical systems, the damping factor ξ is low, so the *peak transmissibility* $T(f_o)$—equal to the normalized displacement x_d/x_o at the resonance frequency f_o in Figures 7.9 and 7.11—is relatively large, on the order of 10–100.

Physically, something which has little dissipation of energy transmits the displacement well, so small ξ implies a high vibration "quality" (symbol Q) associated with a ringing bell.[1] Dissipative damping limits the resonant amplitude but is difficult to quantify. Sources of structural damping include: (1) friction and relative motion between components, (2) metal, plastic, and viscoelastic material deformation with internal friction, and (3) liquid

---

[1] The quality Q of a ringing bell is determined by how long it rings—a good bell rings for a long time, whereas a poor bell damps out quickly with a "thud" sound. This is exactly the opposite of the vibration properties of an optical system: a good system damps out quickly, while a poor one rings for a long time. It is thus an unfortunate circumstance of history that a large Q implies higher quality, when in fact a low Q is preferred for optical systems.

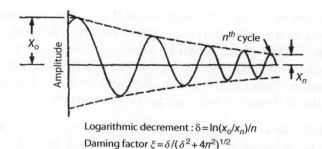

Logarithmic decrement : $\delta = \ln(x_0/x_n)/n$
Daming factor $\xi = \delta/(\delta^2 + 4\pi^2)^{1/2}$

FIGURE 7.12 Damping is measured as a decrease in vibration amplitude for a free vibration. Credit: Ralph M. Richard, "Damping and Vibration Considerations for the Design of Optical Systems in a Launch/Space Environment", Proc. SPIE, Vol. 1340, pp. 82–94.

damping via fluid viscosity, where the damping force is linearly proportional to velocity ($F = cv$).

Fourier analysis tells us that the more the damping slows down the motion— i.e., the bigger the damping constant c—the greater the range of frequencies in the motion, and thus the larger the mechanical bandwidth $\Delta\omega$. Similarly, a small mass has its speed reduced by damping more easily than a heavy mass with large inertia, also resulting in a larger bandwidth. The mechanical bandwidth is thus given by [4]

$$\Delta\omega = \frac{c}{2m} \qquad (7.3)$$

where c is the damping coefficient (units of kg/s) from Figure 7.10 and m is the mass (units of kg). In other areas of physics and engineering, this concept of bandwidth (or *linewidth*) also applies to atomic theory (excited-state lifetime decay), laser resonators (cavity losses), and electrical circuits (resistive losses).

Connecting the mechanical bandwidth with the mass displacement at resonance, we see from Figure 7.9 and 7.11 that the normalized displacement at resonance is approximately given by $x_d(f_0)/x_o(f_0) \equiv T(f_0) \approx 1/2\xi$ for small damping ξ [4]. It can also be shown that the peak transmissibility $T(f_0)$ is the same as the vibration quality factor $Q = \omega_0/2\Delta\omega \approx 1/2\xi$ [4], illustrating that the narrow bandwidth associated with small damping has a large resonant displacement.

For small damping—that is, $\xi \approx 0.2$ or smaller—the peak transmissibility $[T(f_0) = Q]$ and applied acceleration thus determine the maximum dynamic displacement of a sinusoidally vibrating mass

$$x_d \approx x_o \times Q = x_s \times \left(\frac{a}{g}\right) \times Q \qquad (7.4)$$

If either the base (Fig. 7.5) or the mass (Fig. 7.10) is accelerated, the dynamic vibration amplitude x_d at *resonance*—to be added and subtracted from any static displacement given by Equation 7.2—depends on the static displacement x_s, the applied acceleration a (often specified in units of g's, or a/g), and quality factor Q [or *peak* transmissibility $T(f_0)$]. The increase in deflection due to dynamic effects thus has two components:

1. That due to acceleration of the vibration, or a/g. This is simply an increase in apparent weight due to acceleration beyond that due to gravity. For example, a static weight of 1 N in Figure 7.8 vibrated at $5g$'s has a dynamic weight of 5 N.

2. At resonance, an additional term due to the quality factor Q. As we will see in Section 7.3, this is an approximation also valid for fundamental modes of continuous systems such as cantilever beams. This also depends on the damping, with more damping reducing the transmissibility and dynamic deflection near resonance.

The application of the concepts of damping and resonance displacement to the design of an aerial camera is illustrated in Example 7.1.

Example 7.1   As discovered by Von Karman, air flow around a cylindrical aerial camera creates vortex shedding, resulting in a sinusoidal force on the optical system (in the vertical direction in Figure 7.13). What is the deflection (decenter and tilt) of the camera in flight?

To solve this problem, we must answer three questions:

1. What is driving frequency at 160 km/h (100 miles/h)?
2. How does this compare with the natural frequency of the optical system?
3. What is the deflection (decenter and tilt) of the lens?

In addition, if the deflection is excessive, how can it be reduced?

Question 1. Von Karman showed that for air flow around a cylinder such as a telescope tube, the driving frequency f from vortex shedding perpendicular to the cylinder is [3]

$$f = 0.22\frac{v}{D} \quad \text{[Hz]} \tag{7.5}$$

where a larger wind speed v sheds vortices at a higher rate, as does a smaller cylinder diameter D, which allows air to flow over it in less time. For a telescope with a tube diameter $D = 100$ mm (Fig. 7.14), we see that the sinusoidal driving frequency for a 160 km/h relative wind speed is $f = 0.22v/D = 0.22 \times 44.4$ m/s/0.1 m $= 97.7$ Hz.

FIGURE 7.13 Wind flowing from left to right creates vortices to the right of the cylinder and a transverse (vertical) periodic force. Credit: J. P. Den Hartog, *Mechanical Vibrations*, Dover Publications (1985).

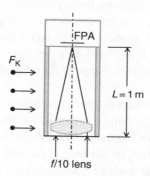

FIGURE 7.14 Schematic of a generic camera bringing light to an image on a focal-plane array (FPA) at the focus of the lens. The wind speed and dimensions of the camera determine its susceptibility to vibration-induced tilt and decenter.

Question 2. As shown in Section 7.3, the lowest natural frequency ω_o of a cantilever beam clamped tightly at one end is

$$\omega_o = 3.52 \sqrt{\frac{EI}{mL^3}} \quad \text{[rad/s]} \tag{7.6}$$

For the cylindrical geometry and dimensions of the camera (inner diameter $D_i = 100$ mm and outer diameter $D_o = 104$ mm), the bending moment of inertia I of the telescope tube is

$$I = \frac{\pi}{64}\left(D_o^4 - D_i^4\right) = \frac{\pi}{64}\left[(0.104 \text{ m})^4 - (0.1 \text{ m})^4\right] = 8.3\text{E}-7 \text{ m}^4$$

Ignoring the mass of the lens at the end of the telescope tube, the aluminum tube's density $\rho_{Al} = 2680 \text{ kg/m}^3$, and its mass m is

$$\begin{aligned} m = \rho V &= \rho \, \pi \left(D_o^2 - D_i^2\right)\frac{L}{4} \\ &= 0.25\pi\left[(0.104 \text{ m})^2 - (0.1 \text{ m})^2\right] \times 1 \text{ m} \times 2680 \text{ kg/m}^3 = 1.72 \text{ kg} \end{aligned}$$

Substituting in Equation 7.6, we find $\omega_o = 642\,\text{rad/s}$, or $f_o = \omega_o/2\pi = 102\,\text{Hz}$— very close to the driving frequency given by Equation 7.5. We thus have a resonance situation where the dynamic amplitude depends on the Q of the system.

Question 3. To find the resonant deflection, the acceleration a is not given, but the applied force F_K is known directly, so the dynamic deflection $x_d \approx x_s Q$ near resonance. That is, because a low-damping system is being driven near a fundamental resonance frequency, its end deflection will be approximated as Q times that due to a non-vibrating force producing a static deflection x_s.

From fluid mechanics, it is known that the drag force $F_K = 0.5\rho v^2\, A_p\, C_d \sin(2\pi ft)$ for a given wind speed v, which has a maximum amplitude when $\sin(2\pi ft) = 1$. As a result, $F_K = 0.5\rho v^2\, A_p\, C_d \approx 0.5\rho v^2\, A_p$ for a drag coefficient $C_d \approx 1$. Also note that the projected area of the tube, given by $A_p = DL$ ($L = 1$ m), is used for the drag force.

It is also known that the lateral force from vortex shedding can be as much as twice the drag force [2]. Using a factor of $1\times$, the maximum driving force is thus $F_K \approx 0.5\rho v^2\, A_p = 0.5 \times 1\,\text{kg/m}^3 \times (44.4\,\text{m/s})^2 \times (0.1\,\text{m} \times 1\,\text{m}) = 98.6\,\text{N}$.

The decenter δ_d and tilt θ_d of the lens at the end of the telescope are then approximated using the methods of Section 5.3 for a uniformly loaded beam with elasticity E and moment I

$$\delta \approx Q\frac{FL^3}{8EI} = \frac{20 \times 98.6\,\text{N} \times (1\,\text{m})^3}{8 \times 69\text{E}9\,\text{N/m}^2 \times 8.3\text{E}-7\ \text{m}^4} = 4.3\ \text{mm}$$

$$\theta \approx Q\frac{FL^2}{6EI} = \frac{20 \times 98.6\ \text{N} \times (1\ \text{m})^2}{6 \times 69\text{E}9\,\text{N/m}^2 \times 8.3\text{E}-7\ \text{m}^4} = 5.8\,\text{mrad}$$

A value of $Q = 20$ was assumed for this lightly damped system; the resulting decenter and tilt are extremely large. Methods to reduce them include stiffening the structure for a higher ω_o via E or I, more damping, and retracting or shrouding the structure so it is not exposed to the airstream.

7.2   RANDOM VIBRATIONS

With the exception of specialized situations such as Karman vortices, "sine" sweeps are not usually found on real platforms. More commonly, vibrations have a range of frequencies and amplitudes. Not including the applied range of frequencies can lead to serious problems in systems which have multiple mass–spring–damper resonances. For example, the system shown in Fig. 7.15 has two resonances which may both be excited at the same time by a random

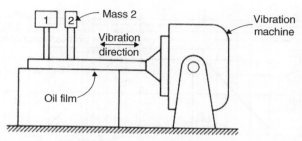

FIGURE 7.15 Systems with more than one natural frequency may result in large, resonance-induced displacements of all masses when subjected to random vibrations. Credit: Dave S. Steinberg, *Vibration Analysis for Electronic Equipment*, John Wiley & Sons (1988).

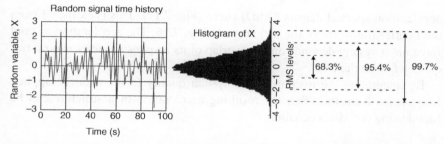

FIGURE 7.16 In the left figure, the acceleration applied to a structure over time is random in both amplitude and frequency. In the right figure, the amplitude distribution is Gaussian. Credit: Keith B. Doyle, Victor L. Genberg, and Gregory J. Michels, *Integrated Optomechanical Analysis* (2nd Edition), SPIE Press (2012).

collection of frequencies. As a result, the vibrations of Mass 1 and Mass 2 may interfere, or even couple through the structure to produce displacements larger than that of either mass analyzed individually.

While sinusoidal vibrations consist of mostly a single frequency, random vibrations contain a collection of many different sinusoids of varying amplitude and frequency (Fig. 7.16). A signal with random amplitudes and frequencies is typically decomposed into its Fourier components for analysis. In structural design, random vibration "amplitudes" are measured as the mechanical power used to shake a structure per unit mass. The resulting units are $F \times v/m = \mathrm{kg\text{-}m/s^2 \times m/s \times 1/kg = m^2/s^3}$. Equivalently, the amplitude can be given as a^2/Hz (or acceleration-squared measured in a 1 Hz bandwidth), or normalized to the gravitational acceleration g as g^2/Hz (pronounced "g-squared per Hertz").

A plot of the mechanical power applied to the structure per unit mass as a function of frequency f is known as a power spectral density (PSD) or

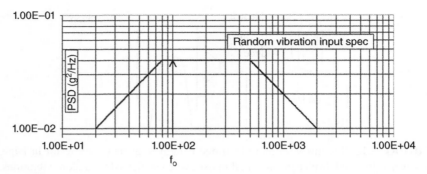

FIGURE 7.17 The power spectral density (PSD) or acceleration spectral density (ASD) curve for a random vibration shows how much mechanical power is applied to a structure over a range of applied frequencies. Credit: NASA (www.nasa.gov).

acceleration spectral density (ASD) curve (Fig. 7.17). This is extracted from the Fourier transform of the $a(t)$ curve in Fig. 7.16. The acceleration of (and force on) a system depends on the overlap of its response near its natural frequency f_o with the PSD.

By integrating the overlap of the system quality factor with the applied frequencies from the PSD, the resulting acceleration of a structure can be found using the Miles equation

$$\frac{a_{norm}^2}{f_n} = \frac{\pi}{2} Q \cdot PSD_n \quad [g^2/Hz] \quad (7.7)$$

where the applied acceleration implicit in the PSD results in a normalized acceleration of the optomechanical system (given by a_{norm}). The normalized acceleration depends on the value of the PSD_n at a given natural frequency f_n; Figure 7.17 illustrates the frequency $f_o = 100\,Hz$ with $PSD_o = 0.04\,g^2/Hz$, a relatively large value.

Equation 7.7 assumes (1) that the PSD_n is approximately constant over the displacement curve and (2) that Q is greater than 10 or so, giving a relatively narrow spectrum over which the applied PSD overlaps (Fig. 7.18). The acceleration a_{norm} is unitless, normalized with respect to the gravitational acceleration g (9.81 m/s²), such that $a = a_{norm} g$. The acceleration a_{norm} is a 1σ root mean square (RMS) value. Typically 3σ (peak-to-valley) values are used for design, with $3a_{norm}$ exceeded only 0.3% of the time [8]; see Figure 7.16.

By integrating the PSD curve, the RMS acceleration applied to the system can also be found. For example, for the PSD plot shown in Figure 7.17, the approximate acceleration is found by integrating the constant PSD over the

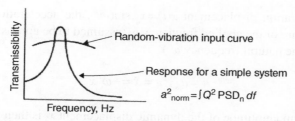

FIGURE 7.18 Equation 7.7 assumes a relatively narrow bandwidth associated with $Q \geq 10$. Credit: Dave S. Steinberg, *Vibration Analysis for Electronic Equipment*, John Wiley & Sons (1988).

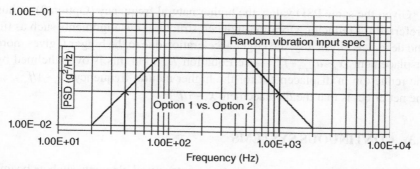

FIGURE 7.19 An optomechanical structure can be designed with a lower or higher natural frequency. Which is preferred? Credit: NASA. www.nasa.gov.

frequency band from 80 to 500 Hz. Thus $a_{norm} = [0.04\,g^2/\text{Hz}° \times (500-80\,\text{Hz})]^{1/2} = 4.1$ (RMS), quoted as "4.1 g's RMS." This result is a *unitless* number in multiple of g's, such that the actual RMS acceleration $a = a_{norm} \times g = 4.1 \times g$ for the above example. These numbers are commonly used, but also commonly misused, in that the number does not tell us how the system will respond to the PSD. For example, it is possible to design a system with its natural frequency outside the band of the PSD, in which case the system will vibrate very weakly. What is needed for the system response, then, is the PSD_n at the value of a natural frequency f_n, as illustrated by the following example.

Example 7.2   When designing a structural system, do we choose lower or higher natural frequencies for reducing the vibration-induced misalignments of tilt, decenter, despace, and defocus? Both options in Figure 7.19 have the same PSD (0.02 g^2/Hz), but Equation 7.7 shows that Option 1 has the smaller acceleration. So which is better from the perspective of dynamic misalignments?

For a vibration displacement $x(t) = x_d \sin(\omega t)$, the acceleration increases with the square of the applied frequency ω (assumed to be approximately resonant with the natural frequency ω_o)

$$a = a_{norm} g = \ddot{x} = -\omega_o^2 x \qquad (7.8)$$

The maximum amplitude of the dynamic displacement x_d is then

$$x_d = \frac{a_{norm} g}{\omega_o^2} = \frac{a_{norm} g}{(2\pi f_o)^2} = g \sqrt{\frac{2Q \cdot PSD_o}{(4\pi f_o)^3}} \qquad (7.9)$$

Given the same PSD value, the higher natural frequency (Option 2) is thus preferred for reducing vibration-induced structural misalignments such as tilt and decenter. Physically, the larger acceleration due to the bigger f_o gives more displacement [$x_d \sim a_{norm} \sim f_o^{1/2}$—see Equation 7.7], but this is overwhelmed by the reduction in displacement for the higher natural frequency ($x_d \sim 1/f_o^2$), so the net effect is that the displacement $x_d \sim 1/f_o^{3/2}$.

7.3 CONTINUOUS SYSTEMS

For optomechanical systems consisting of structural elements such as beams and plates, the number of springs and masses is no longer one or two (Fig. 7.20), but essentially infinite—such systems are called *continuous systems*. With an infinite number of masses (atoms) and springs (interatomic forces), the number of natural frequencies is also infinite. Thus, there is no such thing as "the" natural frequency for continuous systems. Instead, there are usually two or three frequencies that are important, the smallest ("lowest") of which is called the *fundamental frequency*.

An important property of continuous systems is that the vibrating structure takes on a unique shape (eigenmode) for each natural frequency (eigenvalue). This combination of shape and associated frequency is known as a *mode*.

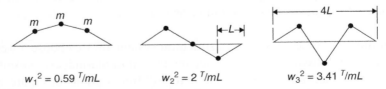

$$w_1^2 = 0.59\ ^T\!/mL \qquad w_2^2 = 2\ ^T\!/mL \qquad w_3^2 = 3.41\ ^T\!/mL$$

FIGURE 7.20 Illustration of the natural frequencies possible for three masses tied together by a string with tension T. Adapted from J. P. Den Hartog, *Mechanical Vibrations*, Dover Publications (1985).

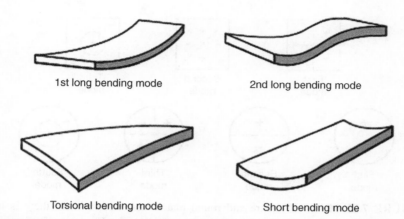

1st long bending mode   2nd long bending mode

Torsional bending mode   Short bending mode

FIGURE 7.21 Short dimensions are stiffer and thus have a higher natural frequency; also see Equation 7.6 and Figure 7.23. Credit: CVI Laser, LLC.

Modes can be of different types—bending or torsional, for example—with both dimensions determining the possible range of shapes (Fig. 7.21). These different types of modes can distort the optomechanical structure holding the optics, creating tilt, decenter, and despace in ways that are not obvious from the static deflection—see Figure 7.21.

Just as in quantum mechanics—the "particle-in-a-box" problem—mode frequencies are based on fitting integer multiples of half-sine waves into a given dimension, subject to the boundary conditions. That is, the size, shape, and boundary conditions determine a mode's natural frequency, so for a clamped guitar string of length L, the resonance condition is $L = n\lambda/2$ for $n = 1$, 2, 3, ... and a mechanical wavelength λ. The wave equation determines the mode shape $y_n(x)$ and vibration frequency ω_n [6].

Many structural elements are not beams—defined by a length at least 5× greater than the width or height—but square and circular plates. Circular plates have the same constraints and boundary conditions as square plates but with mode shapes in cylindrical coordinates (r, θ, z). The *nodes* in Figure 7.22— where the plate's deflection is zero—are shown as dashed lines in the views looking down on the plates.

It is the mode *shape* that determines the natural frequency of a beam or plate, not just the inherent spring stiffness k and mass m. For example, a cantilever's fundamental frequency $\omega_o = 3.52(EI/mL^3)^{1/2}$. From Section 5.3, however, the spring stiffness of a uniformly loaded (self-weighted) cantilever is $k = 8EI/L^3$, from which $(k/m)^{1/2} = 2.83(EI/mL^3)^{1/2}$. The reason for the difference is that the natural frequency and deflections must take into account the Fourier sine- and cosine-shaped components of a vibrating beam, whereas the static deflection of

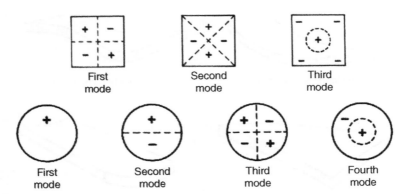

FIGURE 7.22 Modes of square and round plates have nodes where there is no deflection. Credit: Dave S. Steinberg, *Vibration Analysis for Electronic Equipment*, John Wiley & Sons (1988).

a cantilever—while giving approximately correct results for the fundamental mode—does not. Various techniques such as Rayleigh's method and the method of Stadola are available for rigorously calculating the fundamental frequency [3]. A table summarizing the mode shapes and frequencies of beams for various boundary conditions is given in Figure 7.23.[2]

For geometries other than the simple beams shown in Figure 7.23, mode frequencies can often be approximated using handbooks such as *Roark's Formulas for Stress and Strain* [7]. An example is shown in Figure 7.24, where the case of a simply supported uniform beam is highlighted (Case 1b), giving the same natural frequencies as shown in Figure 7.23. Many other cases are given in Ref. [7] as well, including strings, circular plates, etc.

The physical trends evident in Figures 7.23 and 7.24 are that structures can be made stiffer—and the natural frequencies increased—via changes in material and geometry. As we will see in Section 7.4, the material component is governed by the specific stiffness (E/ρ), just as it was for the static bending. For a cylindrical beam, the geometric component is seen through the length L and the D^4-dependence of the bending resistance I. Boundary conditions such as how the beam is supported on its ends—for example, simply supported vs. tightly clamped—also play a role, sometimes as strongly as that of the diameter or overall length.

The modes that are excited in a continuous system depend on the driving frequencies of the vibration sources. For example, random-vibration PSDs

---

[2] Not included in Figure 7.23 are strings (as in a guitar) or membranes (as in a drum or pellicle membrane) whose resonance frequencies depend on the tension used to tighten (or "tune") the string or membrane.

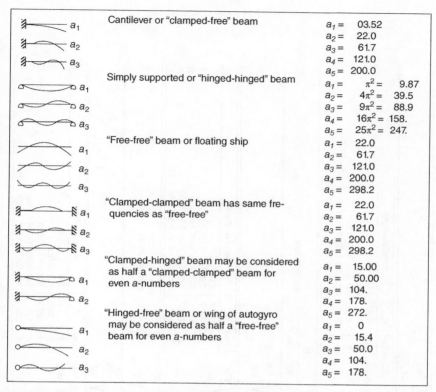

Cantilever or "clamped-free" beam	$a_1 =$	03.52
	$a_2 =$	22.0
	$a_3 =$	61.7
	$a_4 =$	121.0
	$a_5 =$	200.0
Simply supported or "hinged-hinged" beam	$a_1 = \pi^2 =$	9.87
	$a_2 = 4\pi^2 =$	39.5
	$a_3 = 9\pi^2 =$	88.9
	$a_4 = 16\pi^2 =$	158.
	$a_5 = 25\pi^2 =$	247.
"Free-free" beam or floating ship	$a_1 =$	22.0
	$a_2 =$	61.7
	$a_3 =$	121.0
	$a_4 =$	200.0
	$a_5 =$	298.2
"Clamped-clamped" beam has same frequencies as "free-free"	$a_1 =$	22.0
	$a_2 =$	61.7
	$a_3 =$	121.0
	$a_4 =$	200.0
	$a_5 =$	298.2
"Clamped-hinged" beam may be considered as half a "clamped-clamped" beam for even a-numbers	$a_1 =$	15.00
	$a_2 =$	50.00
	$a_3 =$	104.
	$a_4 =$	178.
	$a_5 =$	272.
"Hinged-free" beam or wing of autogyro may be considered as half a "free-free" beam for even a-numbers	$a_1 =$	0
	$a_2 =$	15.4
	$a_3 =$	50.0
	$a_4 =$	104.
	$a_5 =$	178.

FIGURE 7.23 The mode shapes and associated natural frequencies of various beam supports. The resonant frequency $\omega_n = 2\pi f_n = a_n (EI/mL^3)^{1/2}$ in units of radians per second for f_n in units of Hertz (cycles per second). Note that the fundamental frequency in this figure is given the symbol ω_1, not ω_0. Credit: J. P. Den Hartog, *Mechanical Vibrations*, Dover Publications (1985).

may excite many modes at the same time. Low-order modes such as the fundamental require less energy to drive into resonance and thus usually have the largest amplitude when excited by the same PSD value.

As with static structural properties, both the mechanical structure and the optics can deform under vibration loading. The combined vibration of both results in time-dependent optical errors. For example, in the finite-element analysis of a large primary mirror, Genberg and Doyle showed that the random axial vibration of a fused-silica mirror resulted in a frequency-dependent: (1) change in surface power or radius-of-curvature (ΔROC); (2) increase in surface figure, and (3) rigid-body displacement (despace) along the optical axis. Their results are shown in Figure 7.25; note the vibration-induced degradation

Case no. and description	Natural frequencies
1. Uniform beam; both ends simply supported	
1a. Center load W, beam weight negligible	$f_1 = \dfrac{6.93}{2\pi}\sqrt{\dfrac{EIg}{Wl^3}}$
1b. Uniform load ω per unit length including beam weight	$f_n = \dfrac{K_n}{2\pi}\sqrt{\dfrac{EIg}{Wl^4}}$

Mode	K_n	Nodal position/l
1	9.87	0.0 1.00
2	39.5	0.0 0.50 1.00
3	88.8	0.0 0.33 0.67 1.00
4	158	0.0 0.25 0.50 0.75 1.00
5	247	0.0 0.20 0.40 0.60 0.80 1.00

$K_n = n^2\pi^2$

Ref. 22

1c. Uniform load ω per unit length plus a center load W	$f_1 = \dfrac{6.93}{2\pi}\sqrt{\dfrac{EIg}{Wl^3+0.486\omega l^4}}$ approximately
2. Uniform beam; both ends fixed	
2a. Center load W, beam weight negligible	$f_1 = \dfrac{13.86}{2\pi}\sqrt{\dfrac{EIg}{Wl^3}}$
2b. Uniform load ω per unit length including beam weight	$f_n = \dfrac{K_n}{2\pi}\sqrt{\dfrac{EIg}{Wl^4}}$

Mode	K_n	Nodal position/l
1	22.4	0.0 1.00
2	61.7	0.0 0.50 1.00
3	121	0.0 0.36 0.64 1.00
4	200	0.0 0.28 0.50 0.72 1.00
5	299	0.0 0.23 0.41 0.59 0.77 1.00

Ref. 22

2c. Uniform load ω per unit length plus a center load W	$f_1 = \dfrac{13.86}{2\pi}\sqrt{\dfrac{EIg}{Wl^3+0.486\omega l^4}}$ approximately
3. Uniform beam; left end fixed, right end free (cantilever)	
3a. Right end load W, beam weight negligible	$f_1 = \dfrac{1.732}{2\pi}\sqrt{\dfrac{EIg}{Wl^3}}$
3b. Uniform load ω per unit length including beam weight	$f_n = \dfrac{K_n}{2\pi}\sqrt{\dfrac{EIg}{Wl^4}}$

Mode	K_n	Nodal position/l
1	3.52	0.0
2	22.0	0.0 0.783
3	61.7	0.0 0.514 0.868
4	121	0.0 0.358 0.644 0.905
5	200	0.0 0.279 0.500 0.723 0.926

Ref. 22

FIGURE 7.24 Table of beam resonance frequencies. Adapted from *Roark's Formulas for Stress and Strain*, Table 18.1 [7].

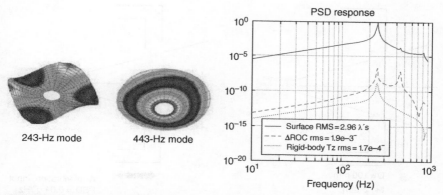

FIGURE 7.25 Modal analysis of a mirror showing the effects on despace ("Rigid-Body"), surface power ("ΔROC rms"), and surface figure ("Surface RMS"). Credit: Keith B. Doyle, Victor L. Genberg, and Gregory J. Michels, *Integrated Optomechanical Analysis* (2nd Edition), SPIE Press (2012).

in ROC, wavefront error (WFE), and despace at the 243-Hz resonance of the mirror. A back-of-the-envelope calculation of WFE for a simple window is illustrated in Example 7.3.

Example 7.3 A vibrating window continuously changes the optical path difference (OPD) and focus of an optical system. The window material, dimensions, and applied PSD are given in Figure 7.26. Is the vibration of the window small enough to maintain diffraction-limited images?

To solve this problem, we must answer three questions:

1. What is the fundamental natural frequency of the window?
2. What is the acceleration and force due to the vibration environment?
3. What is the resulting OPD, and is it acceptable?

Question 1. Table 18.1, Case 10a of *Roark's Formulas for Stress and Strain* shows that the fundamental natural frequency of a circular flat plate with clamped edges is given by [7]

$$\omega_o = 10.21 \sqrt{\frac{Et^3}{12(1-\mu^2)} \frac{A}{mr^4}} = 10.21 \sqrt{\frac{Et^2}{12(1-\mu^2)} \frac{1}{\rho r^4}}$$

$$= 10.21 \sqrt{\frac{104E9 \, \text{Pa} \times (0.003 \, \text{m})^2}{12(1-0.287^2)} \frac{1}{5323 \, \text{kg} / \text{m}^3 \times (0.05 \, \text{m})^4}} = 16,320/\text{s}$$

The window natural frequency in Hz is thus $f_o = \omega_o/2\pi = 2597 \, \text{Hz}$.

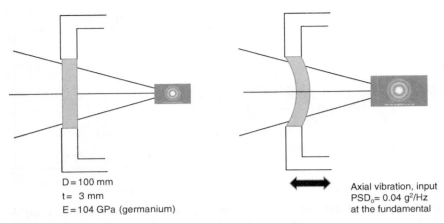

D = 100 mm
t = 3 mm
E = 104 GPa (germanium)

Axial vibration, input
PSD$_o$= 0.04 g^2/Hz
at the fundamental

FIGURE 7.26 Axial vibration of a window dynamically bows the window into a weak lens.

Question 2. We are given an input PSD to the window, not an applied force, so we first calculate the window's RMS (1σ) acceleration

$$\frac{a_{norm}^2}{f_o} = \frac{\pi}{2} Q \cdot PSD_o$$

$$\therefore a_{norm} = \sqrt{\frac{\pi}{2} Q f_o \cdot PSD_o} = \sqrt{\frac{\pi \times 50 \times 2597\,\text{Hz} \times 0.04\,\text{g}^2\,/\,\text{Hz}}{2}} = 90.3$$

The peak window acceleration is thus $a = 3a_{norm}g = 271 \times 9.81\,\text{m/s}^2$. The peak force on the window is then $F = ma = \rho V a$. The peak pressure on the window due to this force is $\Delta P = F/A = \rho ta = 5323\,\text{kg/m}^3 \times 0.003\,\text{m} \times 2658\,\text{m/s}^2 = 42{,}450\,\text{N/m}^2 \approx 6.2\,\text{psi}$.

Question 3. The dynamic pressure thus varies from 0 to 6.2 psi—a very small pressure difference. We can find the OPD from Equation 6.1

$$OPD = 0.01(n-1)\frac{D_w^6}{t^5}\left[\frac{\Delta P}{E}\right]^2$$

$$= 0.01 \times 3 \left[\frac{(0.1\,\text{m})^6}{(0.003)^5}\right]\left[\frac{4.245E4\,\text{Pa}}{104E9\,\text{Pa}}\right]^2 = 0.02\,\mu\text{m}$$

The peak OPD is much less than $\lambda/4$ at any VIS $\rightarrow$ LWIR wavelength, and is therefore not a problem for the given window material, geometry, and vibration conditions.

	Vertical beam 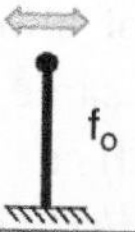 $\quad f_o$	Horizontal beam $\quad f_o$ 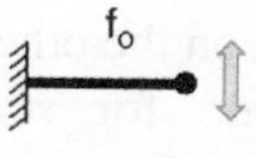
Sinusoidal vibrations 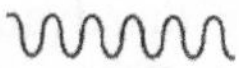	Vibe spec: applied a Output: Max. δ_d at f_o $F = ma$ $\delta_d = Q \times FL^3/8EI$	Vibe Spec: g, applied a Output: Max. δ at f_o $F = m(g+a)$ $\delta = \delta_s + \delta_d = m(g+aQ)L^3/8EI$
Random vibrations 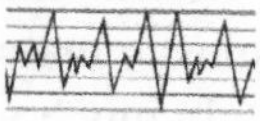	Vibe spec: PSD_o Output: Max. δ_d at f_o $a_{norm} = (0.5\pi Q f_o PSD_o)^{1/2}$ $F = 3ma_{norm}g$ (P–V) $\delta_d = FL^3/8EI$	Vibe spec: PSD_o Output: Max. δ at f_o $a_{norm} = (0.5\pi Q f_o PSD_o)^{1/2}$ $F = mg(1+3a_{norm})$ $\delta = mg(1+3a_{norm})L^3/8EI$

FIGURE 7.27 A comparison of the approximate end deflection at resonance—static gravity offset δ_s plus dynamic bending δ_d—for the fundamental mode of a self-weighted cantilever beam subject to sinusoidal or random vibrations.

Summarizing the results of the first three sections of this chapter, we conclude the following: Figure 7.27 shows a comparison of the approximate end deflection at resonance for the fundamental mode of a continuous system (a self-weighted cantilever beam) subject to either sinusoidal or random vibrations. Two key assumptions made are that (1) the higher-order modes have a relatively small contribution to the dynamic displacement and (2) the initial conditions which determine the dynamic mode shape are based on small static deflections. The figure illustrates the importance of the structural design (via the geometric ratio L^3/I) and materials properties in determining the dynamic deflection. The next section reviews these concepts in more detail.

7.4 STRUCTURAL DESIGN AND MATERIALS SELECTION

As is the case with structural design principles for static loading, the specific stiffness (stiffness-to-density ratio E/ρ) also plays an important role in dynamic loading. For example, the fundamental frequency of a clamped plate can be written in terms of E/ρ as [7]

$$\omega_o = 10.21 \sqrt{\frac{Et^3}{12(1-\mu^2)} \frac{A}{mr^4}} = 10.21 \sqrt{\frac{t^2}{12r^4(1-\mu^2)} \frac{E}{\rho}} \quad \text{[rad/s]} \quad (7.10)$$

Using this approach to increase a structure's natural frequency via materials selection, we find that steel and aluminum are very similar—even though aluminum is lighter by a factor of about 3×. Beryllium and silicon carbide (SiC), on the other hand, offer the same advantages for structural dynamics as they do for static design, with a specific stiffness for beryllium of $303\,\text{GPa}/1850\,\text{kg/m}^3 = 164 \times 10^6\,\text{m}^2/\text{s}^2$—see Table 5.1.

Material damping can control excess ringing, though here the structural materials are limited to a few candidates. Magnesium and cast iron are well known to have excellent damping properties, but both have disadvantages. Cast iron is extremely heavy, for example, and has traditionally been used in high-precision optical-fabrication equipment such as lathes and CNC machines, but is not always appropriate for optical instruments. Magnesium, on the other hand, is very light but also hazardous due to its flammability. More common is to use an elastomer, rubber, or other viscoelastic material to increase damping (Fig. 7.28). These materials are unfortunately much weaker than metals and are thus limited in the weight they can support.

An additional element of structural vibration design is balancing the mass distribution to reduce the moments which create additional vibration-induced motions. "Balancing" requires supporting a system at its center of gravity (CG), where there is no net moment or torque. For example, a uniform beam supported exactly at its CG will not rotate when it is accelerated by vibrations—though it will still bend. A beam supported away from its CG must have mass added to one end to move the system CG to its point of support, so that $F_1 L_1 = F_2 L_2$ (Fig. 7.29). Perfect balancing is not possible, and there will

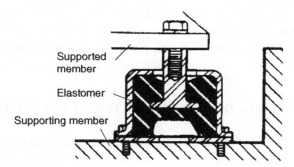

FIGURE 7.28 Elastomeric materials are often used to increase damping. Copyright LORD Corporation 2013.

FIGURE 7.29 Balancing of an optical system reduces the forces and moments which can contribute to vibration-induced motion.

always be some residual imbalance force. In addition, balancing is not always possible for elements within a system, in which case the principles of stiffness and damping must be used.

7.5 VIBRATION ISOLATION

It is not always possible to remove the effects of vibrations with structural stiffening, damping, or balancing. For example, stiffening a structure to meet alignment requirements may require more weight than allowed by the system mass budget. Isolation of the optical system from the vibrations may be necessary in these cases (Fig. 7.30).

We have previously seen how displacements are transmitted by a mass–spring–damper system. Vibration isolation requires reducing the dynamic forces that are transmitted, and the deflections that result. As shown in Figure 7.31 for

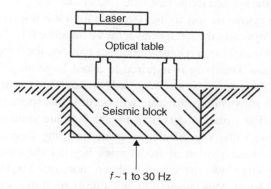

FIGURE 7.30 Base motion of an optical system may require isolation from the source of vibrations with a spring-supported optical table. Credit: CVI Laser, LLC.

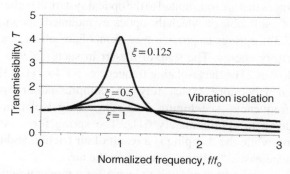

FIGURE 7.31 Transmission curves for base displacement of a simple mass–spring–damper system show vibration isolation (transmissibility $T \equiv x_d/x_o < 1$) at $f > 1.4f_o$.

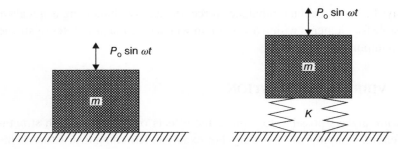

FIGURE 7.32 Isolating an optical system with springs (right) reduces the force acting on the base. Credit: J. P. Den Hartog, *Mechanical Vibrations*, Dover Publications (1985).

a single-resonance system, the displacement transmission curves show attenuation (i.e., the transmissibility $T < 1$) at a cutoff frequency $f > f_0(2)^{1/2}$. Isolating a system from vibrations thus requires lowering the isolator natural frequency f_0 to the point where the applied frequencies are greater than $1.4f_0$.

Mounting the system on soft springs (small k) for a low resonant frequency is often sufficient. Physically, the soft springs act as a trampoline (Fig. 7.32), spreading out the change in velocity over a longer time, thus reducing the force ($F = m\Delta v/\Delta t$) acting on the base. Damping is required to avoid large-amplitude resonant displacement. "Sway space" is also required for the springs to deflect; impacts of the mass with surrounding structure will occur if this space is insufficient.

The base-motion isolation curves in Figure 7.31 are similar to the mass-motion curves with one key difference: more damping *increases* the transmitted force for base motion at frequencies beyond the cutoff frequency. Physically, damping slows the vibration down more quickly, thus increasing the deceleration force. This is generally not a problem if the isolator's natural frequency can be shifted to somewhat lower frequencies to compensate.

Vibration isolation tables are commonly used in the laboratory to minimize the vibration forces that are transmitted to the optical system at higher frequencies. They consist of a stiff table on which the optics are mounted (shown as an inertial mass in Fig. 7.33), springs to provide the "trampoline" effect, and dampers to dissipate vibratory energy. The state-of-the-art in such tables is a resonance frequency as low as 2 Hz, thus isolating frequencies $> 1.4 \times 2\,\text{Hz} \approx 3\,\text{Hz}$.

Commercial laboratory tables rely on compressed-air springs for a low natural frequency—see Figure 7.34—providing isolation for applied frequencies down to approximately 6 Hz. The spring force is supplied by compressed air, while the damping is a result of air friction and the deformation of the viscoeleastic "balloon" containing the air.

Specifications and a transmissibility curve for a moderate-cost laboratory isolator are shown in Figure 7.35. The figure shows isolation against both horizontal and vertical motion; isolation occurs—defined by a transmissibility

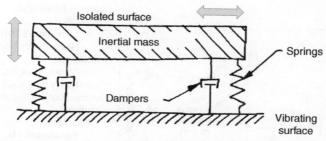

FIGURE 7.33 Schematic of vibration isolation tables typically used in a laboratory environment. Credit: Daniel Vukobratovich, SPIE Short Course SC014.

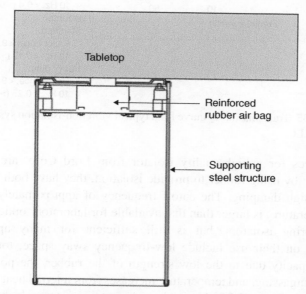

FIGURE 7.34 Hardware schematic of the mass–spring–damper system used in each of the four legs of a common laboratory vibration-isolation table. Credit: CVI Laser, LLC.

$T < 1$—for horizontal frequencies down to approximately 6 Hz. Base motions above that frequency can be reduced by many orders of magnitude, insuring that an optical system mounted on the bench is well isolated from the motions; below that frequency, the base motion is amplified by the isolator, and the optical performance will likely suffer. The transmissibility curve for high-end laboratory isolators shows an even lower resonance frequency, limited by the availability of low-stiffness springs.

Large vibration-isolation tables are generally only useful in a laboratory environment. Smaller commercial off-the-shelf isolators are available for field use from a number of manufacturers (Lord, Stock Drive Products, etc.).

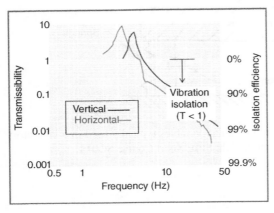

Resonant frequency:
  Vertical:
    Standard: < 5 Hzp;
    Heavy duty: < 5.5 Hz
  Horizontal:
    Standard: < 4 Hz;
    Heavy duty: < 4.5 Hz
Transmissibility (Isolation efficiency):
  Vertical:
    Standard:
      Resonance: < 6
      10 Hz: < 0.25 (> 75%)
    Heavy duty:
      Resonance: < 6.5
      10 Hz: < 0.4 (> 60%)
  Horizontal:
    Standard:
      Resonance: < 9.2
      10 Hz: < 0.13 (> 87%)
    Heavy duty:
      Resonance: < 9
      10 Hz: < 0.25 (> 75%)

FIGURE 7.35 Transmissibility curve for a typical vibration-isolation system. Credit: CVI Laser, LLC.

Typical specs for a high-quality isolator from Lord Corp. are shown in Figure 7.36; by using rubber to provide isolation, they have both low spring force and high damping. The cutoff frequency of approximately 35 Hz (at room temperature) is larger than that available for laboratory optical benches with air-spring isolators, but is still sufficient for many applications. Limitations on their use include low-frequency sway space, total weight-carrying capacity due to the low strength of the rubber, the possibility of excessive outgassing, and temperature increases from internal heat dissipation or, as shown in Figure 7.36, changes in ambient temperature which change the mechanical properties of the rubber—see Problem 7.8.

As shown in Figure 7.37, a vibration isolation system consists of both isolators and an optical bench on which the instrument is mounted. The isolation of an optical system must typically minimize changes in angular pointing of the system, whereas linear displacements may be acceptable—a design complication that requires using matched isolators and placing the isolators at twice the radius of gyration of the bench-plus-optics subsystem [5]. The benches are often constructed using a honeycomb core with significant internal damping to reduce the bench's ringing. The stiffness of the bench—or its inverse, the compliance C—also contributes to the motion and bending of the optical system (Fig. 7.38). As the word implies, a *compliant* bench is not stiff and easily bends or twists under loading.

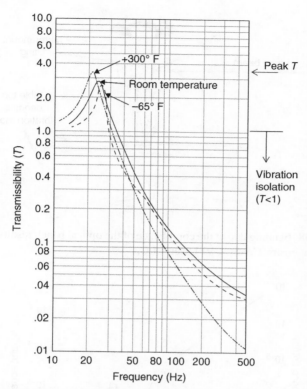

FIGURE 7.36 Typical transmissibility specs for a non-laboratory isolator. Copyright LORD Corporation 2013.

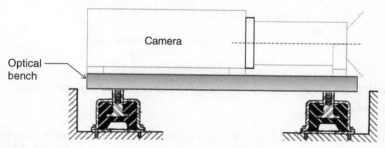

FIGURE 7.37 A camera mounted on an optical bench is isolated from base motions with elastomeric vibration isolators. Isolator drawings: Copyright LORD Corporation 2013.

Compliance C is measured as an inverse spring stiffness (units of mm/N or in/lb). The straight line in Figure 7.39 is the frequency-dependent compliance that is expected for a simple mass–spring–damper system, with $C \sim 1/\omega^2$ [see Eq. 7.8]. The actual compliance curve reveals resonances and anti-resonances

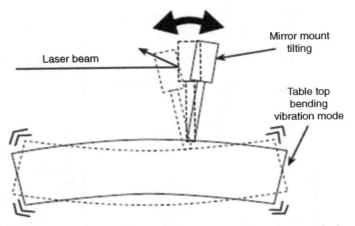

FIGURE 7.38 Bending due to the compliance of the optical bench results in pointing and alignment errors. Credit: CVI Laser, LLC.

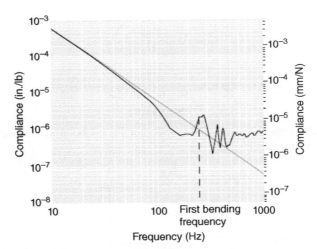

FIGURE 7.39 Compliance curve for a typical laboratory optical table. Credit: CVI Laser, LLC.

due to various modes of the bench. The frequency of the first mode should generally be greater than 100 Hz or so, as determined by the bench's self-loading and stiffness-to-weight ratio. The amplitude should be as low as practical given the constraints of system volume and weight, and the availability of materials for increasing the damping.

Given that the natural frequencies of the optical bench determine the compliance curves, the measured displacement thus depends on the measurement location and the frequency at which the compliance is

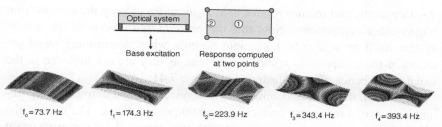

$f_0 = 73.7\,\text{Hz}$ $f_1 = 174.3\,\text{Hz}$ $f_2 = 223.9\,\text{Hz}$ $f_3 = 343.4\,\text{Hz}$ $f_4 = 393.4\,\text{Hz}$

FIGURE 7.40 The first five modes and natural frequencies for the vertical vibration of an optical bench. Credit: Keith B. Doyle, Victor L. Genberg, and Gregory J. Michels, *Integrated Optomechanical Analysis* (2nd Edition), SPIE Press (2012).

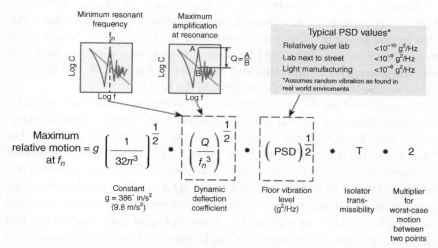

FIGURE 7.41 Summary of the key components contributing to the vertical motion of an isolator-plus-bench system subject to random vibrations. Here, the symbol T is used for the isolator transmissibility, and Q is the quality factor of the optical bench, not the isolator. Permission to use granted by Newport Corporation; all rights reserved.

measured. For example, Figure 7.40 shows an optical bench for a lightweight space-based instrument. Five vertical-vibration modes of the bench are shown, ranging from a simple bending mode (fundamental frequency $f_0 = 73.7\,\text{Hz}$) to a complex "drumhead" mode with a natural frequency $f_4 = 393.4\,\text{Hz}$. As seen in the mode profiles, the displacements at Point 1 and Point 2 are different, depending on the frequency.

As summarized in Figure 7.41, the displacement of a vibration-isolated system depends on the applied PSD, the isolator transmissibility, and the compliance and Q of the bench. However, the displacement of the isolator itself is often a limiting factor in design, as the soft springs used have large displacements, and require room called *sway space* to keep the system from hitting other components. Note that the bending and torsion of the optical system itself must also be taken into account when determining misalignments of the components within the system, as they are not included in the isolator and bench effects illustrated in Figure 7.41.

In general, the transmissibility T of any system can be defined at any frequency where a resonance creates a larger output amplitude. The ratio of the displacement at resonance compared with that expected for a non-resonant system is $T(f_n) = Q$ (given by the amplitude ratio A/B in Fig. 7.41). Also note that Q in Figure 7.41 is the quality factor of the optical bench, not the isolator.

Another term that contributes to the isolator-plus-bench motion in Figure 7.41 is the PSD. As expected from Equation 7.7, the PSD is that value shaking the bench at its resonance frequency f_n. In addition, a 2× worst-case multiplier is used in Figure 7.41 because it is not known where on the bench the instrument may be mounted and, as shown in Figure 7.40, the displacements will be different over the length L of the bench.

As mentioned previously, the isolation of an optical system must typically minimize changes in angular pointing of the system, whereas linear displacements may be acceptable, depending on the available sway space. The "maximum relative motion at f_n" δ given in Figure 7.41 is a first-order displacement which can be used to estimate simple 1D dynamic vibration of a beam (Problem 7.9) and includes the modal bending of the bench, which may result in tilt, decenter, and despace between the optical components mounted on it; including bending such as that shown in Figure 7.40 uses the bending slope δ/L or finite-element analysis—see Chapter 10. Coupling between bench motions—translational and rotational, for example—is also possible; a common method for mitigating such coupling is to place the isolators at twice the radius of gyration of the bench-plus-optics subsystem [5].

7.6 VIBRATION COMPENSATION

It is not always possible to remove the effects of vibrations with stiffening, damping, balancing, or isolation. For example, hand jitter of video cameras cannot be removed with stiffening, and isolation is at best inconvenient. Dynamic motion compensation with an "anti-shake" mechanism for optical image stabilization is required in these cases. In Figure 7.42, for example, a lens is decentered to compensate for angular changes in pointing (tilt) due to hand jitter. In other designs, the FPA is intentionally decentered, also allowing for real-time change in pointing angle to compensate for hand-induced jitter.

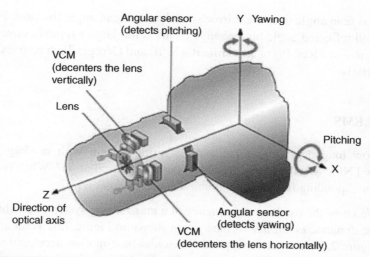

FIGURE 7.42 LOS pointing angle (tilt) is stabilized by decentering a lens in response to an angular sensor. Adapted from Nikon Corp.

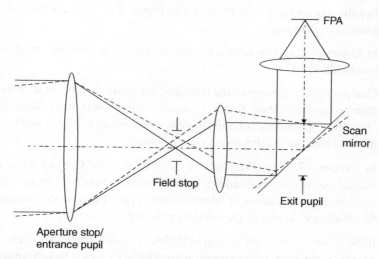

FIGURE 7.43 Small scan mirrors can be placed inside an optical system to compensate for line-of-sight drift and jitter. Adapted from Keith J. Kasunic *Optical Systems Engineering*, McGraw-Hill (2011).

Scan mirrors are also commonly used to compensate for platform motion [9]. Rather than moving the mass of the entire platform itself, a scan mirror allows a lightweight mirror to be moved to stabilize platform LOS motion and jitter (Fig. 7.43). The design trades are between mirror size, mirror bandwidth (scan rate), and scan angle. There is also a factor-of-2 amplification

of optical scan angle over the mirror's mechanical scan angle (because incidence and reflected angle both change). This is too large a topic to cover in
detail here—see Ref. [9] or Jim Hilkert's SPIE and Georgia Tech courses for
more details.

PROBLEMS

7.1 How long does it take for a mass-spring system with $m = 1\,\mathrm{kg}$ and
$k = 1\,\mathrm{N/m}$ to vibrate through one period of oscillation? What is the
corresponding frequency in both rad/s and cycles/s (Hz)?

7.2 We know the oscillation *frequency* of a mass-spring system, but what is
the dynamic *amplitude* of the system shown in Figure 7.6? What about
Figure 7.7? Assume an applied sinusoidal base-motion acceleration of
$2\,g$'s, and very little damping ($Q \sim 100$). Find the amplitude at resonance
($f \sim f_o$) and at very low frequencies ($f \ll f_o$).

7.3 For the mass-spring system shown in Figure 7.6, what are the possible
sources of damping?

7.4 In Equation 7.3, why does a heavier mass give a narrower mechanical
bandwidth?

7.5 Compare the frequency of the fundamental mode for a beam with tightly
clamped ends ("fixed–fixed" boundary conditions) with those using
"free–free" boundary conditions. Which frequency is higher and by how
much? Does the result make physical sense?

7.6 In Example 7.3, a soft elastomer (an "RTV"; see Chapter 9) is used
around the entire circumference of the window to bond it to its mount; the
window has a thickness of 10 mm. What is the *lowest* natural frequency
(approximate, in Hz) of the window-plus-cell system?

7.7 If the window-plus-cell system in Problem 7.6 is vibrated at a frequency
near its resonance, is the vibration amplitude expected to be more, less,
or about the same than if it were driven at a much lower frequency?
Why?

7.8 In Figure 7.36, the isolator's peak frequency is shown to shift to lower
frequencies as the isolator's temperature increases? Does this make
physical sense? Why or why not?

7.9 As shown in the figure below, a laser mirror is tilting due to horizontal
vibrations of the optical bench it is sitting on, causing the beam to

wander randomly. We need to estimate the tilt of the mirror, which depends on (1) the vibration PSD from the platform that the system is mounted on, (2) the transmissibility of the isolator on which the bench is mounted on (not shown), and (3) the resonant frequency of the mirror mount

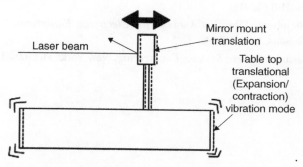

Horizontal vibration of a mirror mount. Credit: CVI Laser, LLC.

Approaching the problem in steps:

a. What is the PSD that is transmitted from the platform through the isolation system (not shown) and into the base of the mirror mount? The PSD of the platform and the transmissibility of the isolation system are given in Figures 7.17 and 7.36, respectively. Note that there is no bench bending in this mode, so the bending compliance of the bench can be ignored.

b. What is the worst-case tilt of the mirror, assuming the mount to be a uniformly loaded cantilever beam? The mount is 150 mm tall, 12.7 mm in diameter, and made of 304 CRES; assume the weight of the mirror is small compared with that of the mount. Note that with almost no friction from air or moving parts in the mount, the resonant Q of the mount is ≈ 100.

REFERENCES

1. Y. Billah and R. H. Scanlan, "Resonance, Tacoma Narrows bridge failure, and undergraduate physics textbooks," Am. J. Phys., Vol. 59, pp. 118–124 (1991).
2. R. G. Fuller, C. R. Lang, and R. H. Lang, *Twin Views of the Tacoma Narrows Bridge Collapse*, College Park: American Association of Physics Teachers (2000).
3. J. P. Den Hartog, *Mechanical Vibrations*, New York: Dover (1985).
4. A. P. French, *Vibrations and Waves*, New York: W. W. Norton & Company (1971).

5. K. B. Doyle, V. L. Genberg, and G. J. Michels, *Integrated Optomechanical Analysis* (2nd Edition), Bellingham: SPIE Press (2012).

6. L. Lyons, *All You Wanted to Know About Mathematics but Were Afraid to Ask*, Vol. 2, Cambridge/New York: Cambridge University Press (1998).

7. W. C. Young and R. G. Budynas, *Roark's Formulas for Stress and Strain*, New York: McGraw-Hill (2001).

8. D. S. Steinberg, *Vibration Analysis for Electronic Equipment*, New York: John Wiley & Sons, Inc. (1988).

9. K. J. Kasunic, *Optical Systems Engineering*, New York: McGraw-Hill (2011).

8

THERMAL DESIGN

Mammals have a sophisticated thermal control system whose sole purpose is to stabilize the temperature of the brain to a predetermined set point—neither too hot nor too cold. To maintain this temperature, nature has adopted a number of clever designs, one of which is illustrated in Figure 8.1. The figure shows that the long ears of rabbits are used for more than sensitivity to sounds; instead, they are also heat-transfer devices. That is, by circulating warm blood through the ears—which have a large surface area in comparison with their volume—the efficient transfer of heat is possible under extreme environmental conditions; this allows the rabbit's core temperature to stay cool in the summer. No doubt other mammals use the same principle as well.

Optomechanical systems commonly use the same concept for maintaining set point temperatures, and many other methods as well. The engineering principles and hardware implementation of these methods are the focus of this chapter. Previously, we saw in Chapters 5–7 the consequences of applied forces such as wind, water, or worse. In this chapter, the forces are not so obvious, but can still have an enormous impact on performance. For example, structural materials expand as they get hotter, and contract as they get cooler, affecting alignments. In addition, differences in expansion between materials—due to differences in temperature or material properties—cause stresses

---

Optomechanical Systems Engineering, First Edition. Keith J. Kasunic.
© 2015 John Wiley & Sons, Inc. Published 2015 by John Wiley & Sons, Inc.

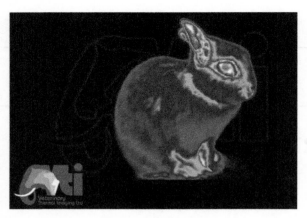

FIGURE 8.1 Infrared images of rabbits illustrate how they keep cool by circulating warm blood through their ears. Long, thin ears have a large surface area in comparison with their volume, facilitating the efficient transfer of heat. Credit: Veterinary Thermal Imaging Ltd.

and deflection in the same manner as an applied load would. These thermal effects are reviewed in Section 8.1.

In addition, there is yet another effect of temperature changes that can be just as large as that of an external force, namely, the refractive index of an optical element typically changes with temperature. As we will see in Section 8.2, this thermal lensing can change the focal length (and WFE) of the element as efficiently as an applied load.

While Section 8.1 and Section 8.2 detail what happens when temperatures change, Section 8.3 reviews how much the temperature will change for a given heat load. Heat can be transferred directly to optical and structural elements either from other elements or from heaters, detectors and FPAs, electronic and mechanical components that dissipate power, and electromechanical components (electrical amplifiers, motors, etc.). Using first-order estimates, we will see that heat-transfer mechanisms known as conduction, convection, and radiation determine how much the temperature changes.

A number of hardware components are available to help us control these changes and their effects, a topic known as *thermal management*. Included are things like heaters, coolers, fins, and fans; Section 8.4 has the details. In addition, judicious materials selection is also a powerful tool available to us for the proper thermal design of an optomechanical system. In Section 8.5, we look at the use of various metrics for material evaluation—including thermal distortion, thermal mass, thermal shock, and the trades between them.

An important note before proceeding to Section 8.1: The term *heat* has in the past been used to mean power (J/s, or watts), power density (W/m^2), or energy (J), with the units and context making clear which meaning is intended. This book uses the "power" definition.

8.1 THERMOSTRUCTURAL DESIGN

Even the simplest of environmental effects—a small change in temperature— can dramatically affect the performance of an optical system. We have seen in Chapter 4, for example, that misalignments such as tilt, decenter, and despace can have critical tolerances—tolerances that can easily be exceeded for small temperature changes with the wrong selection of materials. From the perspective of thermostructural design, the most important material property is the degree to which materials expand or contract when their temperature changes (Fig. 8.2a), a property known as the coefficient of thermal expansion (CTE). Section 8.1.1 reviews this concept, its variations, and its consequences for optomechanical systems engineering.

Even with what may at first glance appear to be the appropriate selection of materials based on CTE, unanticipated temperature differences (or "gradients") can also result in misalignments and distortions (Fig. 8.2b). As shown in Section 8.1.2, these nonuniform gradients result in stresses, and stresses result in strains which alter alignments. Identifying and controlling thermal gradients is thus a key aspect of preventing thermal misalignments, a topic which will occupy much of the chapter.

8.1.1 Thermal Expansion

While there are exceptions such as certain polymers, thermal expansion results from the increased distance individual atoms need to maintain equilibrium as their thermal energy increases (Fig. 8.3). This effect is usually called thermal expansion, even though contraction also occurs as the temperature decreases.

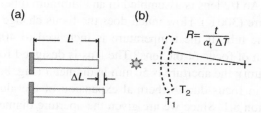

FIGURE 8.2 In (a), a telescope tube expands when uniformly heated, affecting focus and image-quality aberrations; in (b), temperature gradients result in misalignments and stresses.

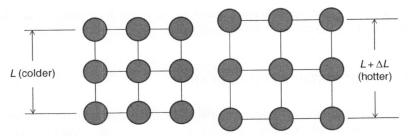

FIGURE 8.3 Microscopic picture of thermal expansion, resulting for most materials in a bigger spacing between atoms at a higher temperature.

On a macroscopic scale, longer lengths have more atoms along a given distance, and so expand more than shorter lengths. To first order, larger temperature changes also cause the material to expand more. Putting things together, the expansion (or contraction) ΔL depends on the length L, the temperature change ΔT, and a material parameter α_t known as the CTE

$$\Delta L = \alpha_t L \Delta T \quad [\mu m] \tag{8.1}$$

Stated differently, the fractional change in length—$\Delta L/L$, or the thermal strain ε_t—is proportional to the CTE and the temperature change, if the CTE is constant over the ΔT

$$\frac{\Delta L}{L} = \varepsilon_t = \alpha_t \Delta T \tag{8.2}$$

The CTE is given the symbol α_t. Typical units are $\Delta L/L$ per unit ΔT, or $1/K$. Because the CTEs are on the order of $10^{-6}/K$, the units are sometimes also written as parts per million (ppm) per degree Kelvin, or ppm/K (see Table 8.1). Low-expansion materials such as Zerodur or Ultra-Low Expansion (ULE) are commonly used for high-stability telescope mirrors, interferometers, and laser cavities.

Example 8.1 An f/2 lens is assembled in an aluminum tube, and focused at room temperature (300 K). How much does the focus change due to thermal expansion of the tube if the temperature is increased to 400 K? Does this exceed the depth of focus of the lens? The lens is designed for visible wavelengths ($\lambda = 0.5\,\mu m$); the aperture is 50 mm in diameter (Fig. 8.4).

The change in focus due to thermal expansion of the aluminum tube is given by Equation 8.1. Since we are given the aperture diameter and the f/#, we can determine the length $L = f$ from $f = f/\# \times D = 2 \times 50\,mm = 100\,mm$. The change in focus due to the expansion of the tube is then given by $\Delta L = \Delta f = \alpha_t f \Delta T = 23.6E\text{-}6/K \times 100\,mm \times (400 - 300)K = 236\,\mu m$.

TABLE 8.1 CTE α_t, Thermal Conductivity k, Thermal Distortion α_t/k, and Transient Thermal Distortion α_t/D for Various Materials Measured at Room Temperature; the CTE is an Average Value Measured Over a Temperature Range of $-30°C$ To $+70°C^a$

Material	CTE α_t (10^{-6}/K)	k (W/m-°K)	α_t/k (10^{-6} m/W)	α_t/D (sec/m²-°K)
Glass Substrates				
Borosilicate (Pyrex)	3.25	1.13	2.88	4.66
ULE	0.03	1.3	0.023	0.04
Zerodur	0.0±0.02 (Class 0)	1.46	0.014	0.03
Refractives				
Fused silica	0.5	1.4	0.36	0.59
Germanium	6.1	58.6	0.10	0.17
N-BK7	7.1	1.11	6.40	13.7
N-SF6	9.03	0.96	9.41	
N-SF11	8.52	0.95	8.97	
Silicon	2.62	140	0.019	0.03
ZnSe	7.1	18	0.39	0.70
ZnS	6.6	16.7	0.39	0.50
Structural				
Aluminum (6061-T6)	23.6	167	0.14	0.35
Beryllium (I-70H)	11.3	216	0.052	0.20
Copper (OFHC)	16.5	391	0.042	0.14
Graphite epoxy	$-1 \rightarrow +1$	3.5	$-0.3 \rightarrow +0.3$	0.05
Invar 36	1.0	10	0.10	0.38
Silicon carbide (Si-SiC)	2.6	155	0.017	0.03
Stainless steel (304)	14.7	16	0.92	3.68
Titanium (6Al-4V)	8.6	7	1.23	3.03

^{a}Data from *Optical System Design* [1], Paquin [2], Weber [3], and *The Crystran Handbook of Infra-Red and Ultra-Violet Optical Materials* [4].

From Chapter 4, the allowable depth of focus (DOF) for an f/2 visible-wavelength lens is $\Delta f = \pm 2\lambda(f/\#)^2 = \pm 2 \times 0.5\,\mu m \times (2)^2 = \pm 4\,\mu m$. The temperature change thus results in a focus change much larger than the DOF. Possible solutions include the use of different tube materials with a smaller CTE (e.g., Invar), a smaller requirement on ΔT, or a focus mechanism. The change in the temperature of the lens itself also has an effect on the focus shift—see Section 8.2.

While a difference in thermal expansion between different materials is often a detriment, it can sometimes be used to advantage. An example is shown in Figure 8.5, where a difference in CTE is used to compensate for the loss of focus due to the expansion of individual materials. The concept is that

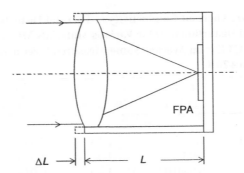

FIGURE 8.4 Thermal expansion along the length of a structure is in part responsible for the change in focus of an optical system as its temperature changes. Adapted from Keith J. Kasunic, *Optical Systems Engineering*, McGraw-Hill (2011).

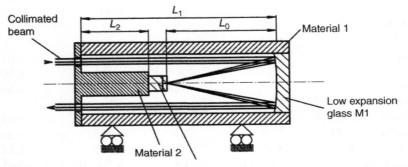

FIGURE 8.5 Thermal expansion of Material 1 away from the primary mirror M1 is compensated by an equal expansion in the opposite direction of Material 2. Credit: P. Giesen and E. Folgering, "Design guidelines for thermal stability in opto-mechanical instruments," Proc. SPIE, Vol. 5176.

one material expands in one direction, while the second material expands an equal amount in the opposite direction—an architecture sometimes known as "re-entrant." The spacing between the primary and secondary mirror is thus maintained; this requires that the longer material have a smaller CTE than the shorter, such that $\Delta L = \alpha_1 L_1 - \alpha_2 L_2 = 0$ for $\alpha_1 < \alpha_2$ [5]. If lenses are also used, however, such designs must take into account the change in lens index with temperature (see Section 8.2).

The discussion to this point has assumed that the CTE is constant over temperature; in practice, this is not the case. In addition, the CTE is not constant within a material, and the temperature change itself may not be uniform (thermal gradients).

When the CTE is not constant over temperature—in the design of cryogenic instruments with large temperature changes, for example—the CTE curve must be integrated to obtain the change in length due to thermal expansion and contraction. This is illustrated in Figure 8.6.

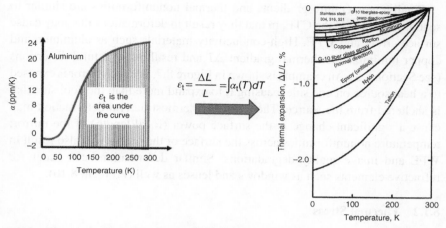

FIGURE 8.6 The CTE of a material depends on its temperature, generally reaching a constant when the material cannot shrink or expand any further. Credits: (left) Keith B. Doyle, Victor L. Genberg, and Gregory J. Michels, *Integrated Optomechanical Analysis* (2nd Edition), SPIE Press (2012); (right) P. Thomas Blotter and J. Clair Batty, "Thermal and Mechanical Design of Cryogenic Cooling Systems", *The Infrared and Electro-Optical Systems Handbook*, Vol. 3, Chap. 6, SPIE Press (1993).

TABLE 8.2 CTE Inhomogeneities within a Material can Determine Thermal Distortions and Wavefront Error[a]

Material	CTE α_t (ppm/K)	$\Delta \alpha_t$ (ppb/K)
Aluminum (6061-T6)	23	60
Beryllium (I-70A)	11.5	30
Borosilicate glass	3.2	30
Fused silica (7940)	0.56	2.0
ULE (7971)	0.03	4.0
Zerodur	0.05	40

[a]Data from Dan Vukobratovich, "Optomechanical Design," *The Infrared and Electro-Optical Systems Handbook*, Vol. 4, Chap. 3, SPIE Press (1993).

At any given temperature, the CTE is also not constant within a material, but has a spatial inhomogeneity $\Delta \alpha$ which can cause surface errors and WFE, even if the temperature distribution is completely uniform. As shown in Table 8.2, these inhomogeneities at room temperature are on the order of parts per *billion* (ppb), or approximately 1000 times smaller than the CTEs themselves for common materials. Nevertheless, they can be a driving factor in the design of high-precision mirrors that use low-expansion and low-inhomogeneity substrates such as ULE.

Finally, temperature gradients and thermal nonuniformities are similar to spatial variations in the CTE, in that they result in deformations that may cause surface errors and WFE. High-conductivity materials such as aluminum and copper minimize the thermal gradient ΔT and resulting structural distortions (see Section 8.5). An example is shown in Figure 8.7, where a mirror is exposed to a heat source on one side, causing it to expand more than the cool side that is sheltered from the source. The resulting thermostructural deformation may cause a significant change in the surface power (focal length) of the mirror; temperature nonuniformities across the surface of the mirror will also result in WFE and image-quality degradations. Similar deformations can occur for refractive elements such as windows and lenses as well (Problem 8.10).

8.1.2 Thermal Stress

What happens if a material is heated, but prevented from expanding? As shown in Figure 8.8, a force F (and stress σ) needs to be applied to the beam

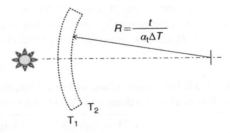

FIGURE 8.7 A temperature difference (or gradient) $\Delta T = T_1 - T_2$ along the axis of a flat mirror causes it to bend into a curved shape with reflective power $\varphi = 1/f = 2/R$ given the CTE α_t and mirror thickness t.

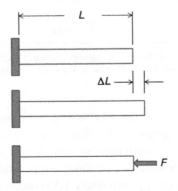

FIGURE 8.8 A stress known as a thermal stress can be applied to a structure to keep it from expanding or contracting as its temperature changes.

 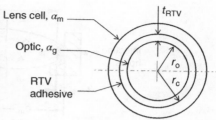

FIGURE 8.9 Differential thermal expansion ($\Delta\alpha_t = \alpha_m - \alpha_g$) between an optic and its lens cell can result in thermal stresses which exceed the strength of the optic. Photo credit: P. Giesen and E. Folgering, "Design guidelines for thermal stability in opto-mechanical instruments," Proc. SPIE, Vol. 5176.

to keep it from expanding (or to force it back to its original size after expanding). The applied stress needed to constrain the beam is known as a *thermal stress*. The thermal stress σ_t depends on the temperature change ΔT, CTE α_t, and the elastic modulus E of the material, such that $\sigma_t = F/A = \varepsilon_t E = \alpha_t \Delta T \times E$. These thermal stresses are just as effective at straining or breaking a lens as the applied loads in Chapters 5–7.

Differences in CTEs can also result in thermal stresses. For example, if a cooled lens cell contracts down on a lens more than the lens contracts away from the cell, the cell applies a force and a stress. This occurs if the CTE of the lens cell is greater than that of the lens (Fig. 8.9). So the stress results from a difference in expansion based on the CTE mismatch (or $\Delta\alpha_t$) between two different ("dissimilar") materials, or a difference in temperature between the lens and cell, or a combination of both.

A simple example of a fused silica lens in an aluminum mount illustrates the difficulties of stress-free mounting for optics in a changing thermal environment. From Table 8.1, the difference in CTE is $\Delta\alpha_t = \alpha_m - \alpha_g = (23.6 - 0.5) \times 10^{-6}/\text{K} = 23.1\,\text{ppm/K}$, so the aluminum is shrinking down on the lens diameter if they are uniformly cooled. Assuming they are assembled with near-zero clearance, the stress in the lens will depend on the elasticity of both the lens and the aluminum cell. For a reasonable ratio of the lens diameter to the lens–cell wall thickness (e.g., 10:1), the thermal stress in the lens is *approximately* $\sigma = \varepsilon_t E_m/10 = \Delta\alpha_t \Delta T \times E_m/10 = 23.1 \times 10^{-6}/\text{K} \times 65\,\text{K} \times 68.9\,\text{GPa}/10 = 10.3\,\text{MPa}$ ($\approx 1500\,\text{psi}$) for a lens which is cooled down to $-40\,\text{C}$ from $+25\,\text{C}$ (see Table 5.1 for the elastic modulus of aluminum). Compared with typical tensile glass strength of 7 MPa or less, the stress in the fused silica is approaching its compressive strength—a guaranteed way to "crunch" an optic.

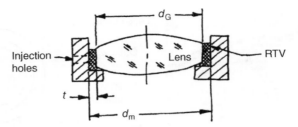

FIGURE 8.10 A stress-free mount over a wide temperature range can be designed using a compliant adhesive (RTV) between the lens and its cell. Credit: Daniel Vukobratovich, SPIE Short Course SC014.

Fracture can be avoided if the lens and cell are assembled with nonzero clearance. This is acceptable if the operating temperature is known and fixed—such as the optics in a cryogenic system—or if the decenter tolerance allows a large enough clearance to accommodate the temperature changes. If the operating temperature varies or the decenter is excessive, however, then a more flexible approach is needed. Such "flexibility" is provided by a room temperature vulcanizing (RTV) adhesive, a relatively compliant material in comparison with the lens and cell materials.

Figure 8.10 shows a design for a stress-free mount over any temperature range where the CTEs of the lens, cell, and RTV are constant. RTV566 is a good material for such applications.

The optimum thickness t of the RTV required for a stress-free lens depends on the CTEs (α_r = CTE of the RTV) and the diameter of the lens D_g [9][1]

$$t = \frac{D_g}{2}\left(\frac{\alpha_m - \alpha_g}{\alpha_r - \alpha_m}\right) \quad \text{[mm]} \tag{8.3}$$

Even with such an optimum RTV thickness, there will still be unequal cooling rates during heat-up or cool-down leading to temperature gradients and thermal stresses. For example, it is common in cryogenic systems for the lens cell to cool down (and contract) before the lens, as it is more directly connected to the cooling source. Equation 8.3 does not include this non-equilibrium effect, and finite-element analysis (FEA) is generally needed for this level of detail (see Chapter 10).

---

[1] Note that if the CTEs of the metal and the glass are equal, then the bond thickness is zero, a result which is clearly outside the bounds of the validity of the equation.

8.2   THERMO-OPTIC AND STRESS-OPTIC EFFECTS

What happens when a lens or window gets hotter or cooler? There are three primary effects:

- As with the structural elements in Section 8.1, the lens expands or contracts, which changes the thickness and surface power, and may also create wavefront error (WFE).
- The lens refractive index changes (thermo-optic effect, dn/dT), which also creates WFE.
- The thermal expansion and contraction may also result in thermal stresses, which change the lens refractive index (stress-optic effect, $dn/d\sigma$) and create WFE.

The expansion or contraction of a lens changes its surface radii but does not necessarily create aberration WFE. A *uniform* temperature change across an unconstrained lens, for example, expands the lens uniformly with defocus WFE. If the refractive index does not also change with temperature, the effective focal length also increases ($\Delta f = \alpha_t f \Delta T$) from the change in lens radii ΔR. As shown in Figure 8.11, this effect is similar to making a photocopy of the lens at a magnification that depends on the CTE and the temperature change— mirror systems can be designed this way if the ΔT is uniform.

In practice, it is almost never the case that the temperature changes and distribution in a complex system are uniform. More commonly, the lens

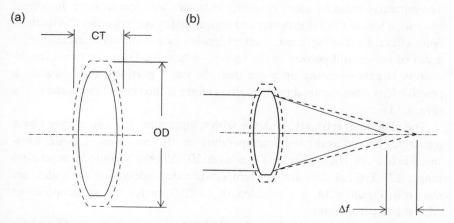

FIGURE 8.11 (a) An unconstrained lens expands uniformly when heated; (b) if the lens index does not change, this increases the focal length due to the increase in radii of the lens surfaces.

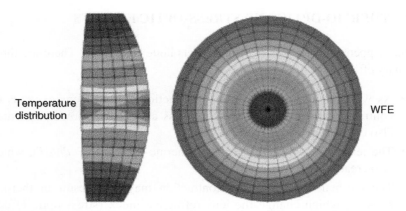

FIGURE 8.12 Nonuniform temperature distributions in an optical element create WFE due to the thermal expansion and the change in index with temperature (dn/dT). Credit: Keith B. Doyle, Victor L. Genberg, and Gregory J. Michels, *Integrated Optomechanical Analysis* (2nd Edition), SPIE Press (2012).

temperature will not be uniform, but instead have a spatial profile across the surface that creates WFE (Fig. 8.12). Such WFE is typically predicted numerically using FEA. As shown in later sections, there are two simplified cases where the defocus WFE can be predicted analytically: radial and axial temperature gradients resulting from the thermo-optic effect (*dn/dT*).

8.2.1 Thermo-Optic Effect

The refractive index of a lens generally increases with temperature. In optical systems, a loss of resolving power and image quality accompanies this thermo-optic effect. In laser systems, thermal lensing is a common consequence, a result of the central portion of the optics—where the Gaussian beam irradiance is largest—heating up more than the outer portions, thus creating a positive lens whose optical path length is larger at the center than at the edges (Fig. 8.13).

The change in refractive index n with temperature T is summarized by a parameter known as the index temperature coefficient (symbol *dn/dT*, measured in units of 1/K), where $n(T) \approx n_o + (dn/dT)\Delta T$ over a limited temperature range ΔT. Typical data for the temperature dependence of the index are shown in Figure 8.14 at a wavelength $\lambda = 10.6\,\mu m$ for the semiconductors silicon and germanium.

For common optical glasses such as N-BK7 (or equivalent), *dn/dT* data are available from SCHOTT and other glass manufacturers. As shown in Table 8.3, the *dn/dT* (or $\Delta n/\Delta T$) values are both temperature and wavelength dependent. The uncorrected value (dn_{abs}/dT) should be used for optical systems that

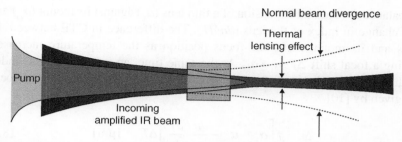

FIGURE 8.13 Thermal lensing from a nonuniform temperature distribution in a laser crystal creates a positive lens, slightly focusing the beam. Credit: Coherent, Inc.

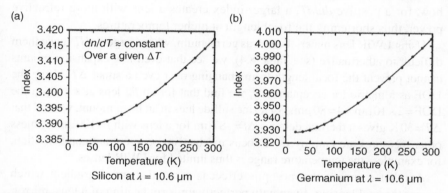

FIGURE 8.14 The refractive index for many optical materials increases with temperature, as shown here for (a) silicon and (b) germanium. Credit: J. M. Hoffman and W. L. Wolfe, "Cryogenic Refractive Indices of ZnSe, Ge, and Si at 10.6 μm," Applied Optics, 30(28), pp. 4014–4016 (1991).

TABLE 8.3 The Change in Index with Temperature ($\Delta n/\Delta T$) Depends on Both Temperature and Wavelength[a]

Temperature range (C)	$\Delta n_{rel}/\Delta T$ (10⁻⁶/K)			$\Delta n_{abs}/\Delta T$ (10⁻⁶/K)		
	1060 nm	546.1 nm	435.8 nm	1060 nm	546.1 nm	435.8 nm
$-40 \rightarrow +20$	2.4	2.9	3.3	0.3	0.8	1.2
$+20 \rightarrow +40$	2.4	3.0	3.5	1.1	1.6	2.1
$+60 \rightarrow +80$	2.5	3.1	3.7	1.5	2.1	2.7

[a] Data from SCHOTT N-BK7 data sheet

operate in a vacuum, where the change in the refractive index of air with temperature does not contribute to the total dn/dT.

One application of the thermo-optic concept is the design of an *athermal* lens whose focus position does not change with temperature. The approach is to balance the thermal expansion of a thin lens (α_L) against its mount (α_m) and the change in index of the lens (dn/dT). The difference in CTE between the lens and its mount shifts the focus position as the temperature increases, giving a focal shift $\Delta f \sim \alpha_L - \alpha_m$. At the same time, an increase in lens index with temperature decreases the focal length [$1/f \sim (n-1)/R$], with the net defocus Δf given by [10]

$$\Delta f = f\left[\alpha_L - \alpha_m - \frac{dn/dT}{n-1}\right]\Delta T \quad [\mu m] \qquad (8.4)$$

Note that the negative sign for the dn/dT term in Equation 8.4 illustrates how, for a positive dn/dT, a larger index creates a lens with more refractive power, thus shortening the focal length at higher temperatures.

Using LWIR lens materials such as germanium, whose large dn/dT make them difficult to athermalize (see Table 8.4), we see that even low-expansion mounts cannot prevent the focal length from changing over even a small ΔT. Using the DOF as a metric for acceptable Δf, we find that for an $f/2$ lens at $\lambda = 10\,\mu m$ the DOF $= 2 \times 10\,\mu m \times 4 = 80\,\mu m$. For a zinc sulfide lens in an Invar mount, we find that $\Delta T = 30\,K$ gives a thermal defocus $\Delta f = -87\,\mu m$ for a lens with $f = 100\,mm$. Unless other measures are employed—a focus mechanism or re-entrant compensation, for example—the temperature range is thus limited for LWIR lenses.

Another common thermo-optic effect is a radial thermal gradient which causes thermal lensing. Even with perfectly uniform heating of a lens, mirror,

TABLE 8.4 Relative Change in Index with Temperature (dn_{rel}/dT) for Various Refractive Materials Measured in Air Over the Temperature Range 20–40°C at the Wavelengths Indicated [11]^{a,b}

Material	Wavelength	dn/dT (10^{-6}/K)	$dn/d\sigma$ (10^{-12}/Pa)
AMTIR 1	LWIR	72	
Germanium	LWIR	385	
N-BK7	541.6 nm	3.0	2.77
N-SF6	541.6 nm	1.5	2.82
N-SF11	541.6 nm	2.4	2.94
Silicon	MWIR	162	
Zinc Selenide	LWIR	61	
Zinc Sulfide	LWIR	45	

Also given is the change in index with stress ($dn/d\sigma$). Values supplied by different references vary, especially for IR materials.
^{a}The dn/dT for the IR materials were calculated from the CTEs listed in Table 8.1 and the refractive index n and thermo-optical constant data given in Bass et al. (1995).
^{b}Data are from Rogers and Roberts [10], Smith [12], and see the SCHOTT glass website for more details (www.schott.com).

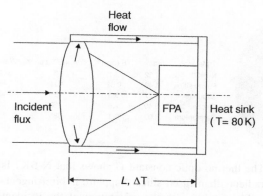

FIGURE 8.15 Radial thermal gradients can easily result from the heat transfer of an absorbed incident flux. Adapted from Keith J. Kasunic, *Optical Systems Engineering*, McGraw-Hill (2011).

or window, the transfer of heat to or from the optic can lead to these gradients. Figure 8.15 illustrates the situation where the center of a lens is hotter—and therefore has a higher index than the edges—due to the transfer of heat (absorbed flux) out through the edges. Thermal lensing is even more noticeable with laser beams with a Gaussian irradiance profile, leading to additional heating of the center of the lens where the laser irradiance is largest.

In such situations, a radial thermal gradient ΔT_r produces a defocus WFE given by [13]

$$\text{WFE} = t\left[\alpha_t \left(n-1\right) + \frac{dn}{dT}\right]\Delta T_r \quad [\mu\text{m}] \qquad (8.5)$$

This equation shows that this peak-to-valley WFE has two components: a change in lens thickness t due to its thermal expansion $\Delta t = \alpha_t t \Delta T_r$ and that due to a thermo-optic index change dn/dT. This leads to the definition of the thermo-optic constant $G = \alpha_t(n-1) + dn/dT$, used as a metric for sensitivity to radial gradients (where small G is better—see Figure 8.16 for typical values at room temperature). This does not mean that the least sensitive material is the best option, as other factors such as material absorption determine the radial temperature gradient ΔT_r itself; methods for estimating ΔT_r will be reviewed in Section 8.3.

8.2.2 Stress-Optic Effect

A temperature change may result in radial thermal stresses which, from Poisson's ratio, cause deformation and WFE of the surfaces (Fig. 8.17). These

Material	Index	CTE (ppm/K)	dn_{rel}/dT (ppm/K)	Wavelength	G (ppm/K)
N-BK7	1.52	7.1	3	541.6 nm	6.7
Fused silica	1.46	0.52	10.2	546.1 nm	10.4
Sapphire	1.77	5.3	13.1	546.1 nm	17.2
Silicon	3.40	2.6	162	MWIR	168.2
Germanium	4.00	6.1	385	LWIR	403.3

FIGURE 8.16 The thermo-optic constant G shows that N-BK7 is less sensitive to radial thermal gradients than fused silica. Sapphire is birefringent—that is, it has a refractive index which varies with the orientation of the incident electromagnetic field—but the value for G listed here is representative. Graphic credit: Keith B. Doyle, Victor L. Genberg, and Gregory J. Michels, *Integrated Optomechanical Analysis* (2nd Edition), SPIE Press (2012).

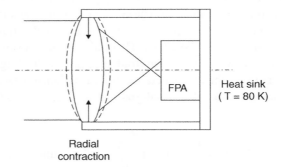

FIGURE 8.17 Differences in CTE may result in compression of the lens diameter, leading to expansion of the thickness. Adapted from Keith J.Kasunic *Optical Systems Engineering*, McGraw-Hill (2011).

stresses may also change the lens refractive index via the stress-optic effect (symbol $dn/d\sigma$), though it is not necessary for the stresses to be of a thermal cause, and could just as easily be the result of an applied load such as those reviewed in Chapters 5–7.

The magnitude of the stress-optic effect is usually small. For an upper-limit stress based on a nominal strength $\sigma \approx 10$ MPa, Table 8.4 shows that the stress-induced index change Δn for a typical glass with $dn/d\sigma \approx 3 \times 10^{-12}$/Pa is given by $\Delta n = \sigma \times dn/d\sigma = 10$ MPa $\times$ 3E-12/Pa $= 30 \times 10^{-6}$ (30 ppm). The change in index is thus negligible for most classes of imaging sensors, but could be significant for phase-sensitive instruments such as lasers or interferometers.

8.3 HEAT TRANSFER

In the previous section, we saw that a radial thermal gradient modifies the WFE of a lens. In addition, identifying and controlling thermal gradients is a key aspect of preventing thermal misalignments. Not yet covered, however, are the tools needed to estimate the magnitude of the thermal gradient. More generally, predicting how hot an optical system gets, as well as how to design the system to manage its temperature, are the goals of this section.

The sources of heat in optical systems are many. They might include the following:

- Electronics and motors
- Solar loading
- Absorption of laser pump sources or output power
- Friction (e.g., from moving mechanisms)
- Ambient temperature (for cooled systems)
- Wall-plug inefficiency of optical sources (lasers, LEDs, etc.)

In all cases, the temperature of the system increases—including individual optical or mechanical components—until the heat transferred away equals the heat created by the source. The temperature rise thus depends on both the heat load and the thermal resistance to heat transfer. As a result, the system may need to get very hot to transfer the heat, and reducing the barriers to heat flow is the key to the thermal management of the temperature rise.

As one example of the importance of heat transfer, the conversion of pump energy to laser output energy has a number of thermal inefficiencies. The output power of many lasers is thus limited by the amount of waste heat generated, as well as the ability to efficiently remove this heat (Fig. 8.18).

Three mechanisms transfer heat away from any system: conduction, convection, and radiation. Conduction and convection are based on the physical contact of cold atoms with warm. Radiation (e.g., solar), on the other hand, does not require contact, and can occur across a vacuum. For all three mechanisms, a temperature difference is required for a *net* transfer of heat. Details are given in the following sections.

8.3.1 Conduction

The most noticeable form of heat transfer is conduction, where warm atoms transfer their lattice vibrational energy to cold atoms via direct contact; neither the hot or cold atoms are moving on a macroscopic scale. Conduction is described using the concept of *thermal resistance*. Thermal resistance

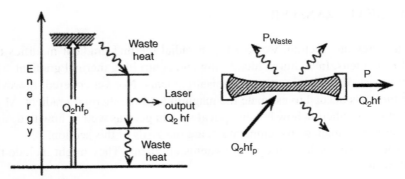

FIGURE 8.18 Nonradiative atomic transitions which do not produce photons produce waste heat instead. Credit: Hugo Weichel, *The Infrared and Electro-Optical Systems Handbook* [14].

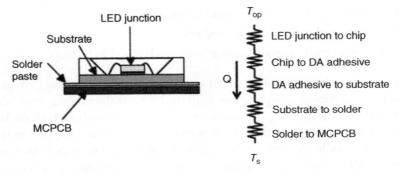

FIGURE 8.19 A component with a large thermal resistance must get very hot to remove a watt of power dissipated from a component such as a high-brightness LED. Adapted from Avago Technologies.

determines how hot something will get for a given amount of heat transferred. As shown in Figure 8.19, a large thermal resistance produces a large temperature difference when transferring one watt of heat ($\Delta T = QR_t$), in the same way that a large electrical resistance produces a large potential difference when conducting one amp of current ($\Delta V = iR$).

Quantitatively, Fourier's law of heat conduction describes the temperature rise expected for a given heat load Q and thermal resistance R_t. In its engineering form with the thermal resistance independent of Q, Fourier's law shows that how hot a component must get to remove heat—its temperature rise ΔT—is directly proportional to both the heat load and the thermal resistance

$$\Delta T = T_{op} - T_s = R_t Q \quad [\text{K}] \tag{8.6}$$

The thermal resistance R_t for conduction depends on the material and geometry of the path to the heat sink. For the linear geometry shown in Figure 8.19, which consists of heat conducting from an LED junction to the heat sink at a temperature T_s, the thermal resistance is given by [15][2]

$$R_t = \frac{L}{kA} \quad [\text{K / W}] \tag{8.7}$$

For conduction, the thermal resistance depends on the thickness or length L across which heat is being transferred, the cross-sectional area A, and a material property known as the thermal conductivity k. Physically, a longer length exhibits more thermal resistance—a sort of "inefficiency" as heat is transferred from atom to atom—as does a smaller cross sectional area, a result of fewer atoms being available to transfer heat.

The thermal resistance, with units of K/W, is thus a direct measure of the temperature difference created per watt of transferred heat. A large thermal resistance can be good or bad—it depends on the requirements of the system. For example, when trying to keep something from getting too hot—high-brightness LEDs, for example—a small thermal resistance is needed. But when trying to refrigerate something to cooler than room temperature—an IR detector, for example—a large thermal resistance to insulate the system reduces the amount of power needed to maintain the cold ΔT. The three contributors to conductive thermal resistance—L, A, and k—are described in more detail in the following sections. In addition, a fourth contributor known as an interface resistance is included.

8.3.1.1 Thermal Conductivity

The material component of the thermal resistance is the thermal conductivity k (units of W/m-K; see Table 8.1); note that some materials are more efficient than others at transferring heat, which means they have greater thermal conductivity. Quantitatively, thermal conductivity is the number of W/m² needed to produce a temperature difference of 1 K along a length of 1 m. Inversely, a large (small) conductivity results in a small (large) temperature difference per W/m² of heat conducted.

Thermal conductivity can increase or decrease with temperature. For example, the thermal conductivity of aluminum decreases from room temperature by a factor of approximately 10× at cryogenic temperatures (Fig. 8.20a). The thermal conductivity of copper, on the other hand, increases from room temperature by a factor of approximately 50× at 10 K (Fig. 8.20b). For these large temperature changes—as are common in cryogenic systems—the heat

---

[2] Thermal resistance equations for specialized geometries can be found in Refs. [15] and [17].

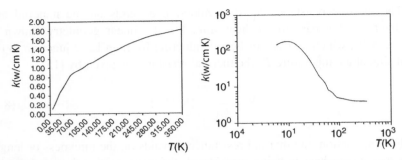

FIGURE 8.20 The thermal conductivity k for (a) aluminum and (b) copper depends on temperature. Credit: P. Thomas Blotter and J. Clair Batty, "Thermal and Mechanical Design of Cryogenic Cooling Systems," *The Infrared and Electro-Optical Systems Handbook*, Vol. 3, Chap. 6, SPIE Press (1993).

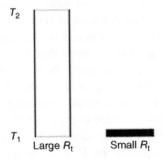

FIGURE 8.21 Both the cross-sectional area and the material length determine the thermal resistance. The tube on the left has a larger thermal resistance R_t—and larger ΔT per watt of heat transferred—because of its long length and small cross-sectional area.

transferred cannot be determined from Equations 8.6 and 8.7, but instead depends on the integrated thermal conductivity [7].

The thermal conductivity also affects thermal stresses, in that large temperature gradients will be created for a small thermal conductivity k. Large gradients in conjunction with a large thermal expansion α_t increases strain, surface distortion (WFE), and stress. The ratio of α_t/k (units of m/W) is thus used as a thermostructural metric, with a smaller ratio preferred—see Table 8.1 and Section 8.5.2 for a comparison of different materials.

8.3.1.2 Cross-Sectional Area
In addition to thermal conductivity, the thermal resistance also depends on the cross-sectional area that is conducting heat to the sink (Fig. 8.21). A larger area has more atoms to transfer heat, in direct proportion to the number of atoms. The thermal resistance thus varies inversely with area (~1/A).

A small area thus has a larger ΔT for a given amount of heat to be transferred. Equivalently, a small area transfers less heat (insulates better) for a given ΔT.

8.3.1.3 Path Length

In addition to thermal conductivity and cross-sectional area, the thermal resistance also depends on the length of material that is conducting heat (Fig. 8.21). Transferring energy from hot to cold atoms has an inefficiency, so the more atoms that are in the conduction chain, the bigger the ΔT along the length. For example, microbolometers measure LWIR radiation by heating up a thin α-silicon or vanadium oxide (VOx) film. How much the temperature increases—and hence the signal level—depends on minimizing the heat lost to its surroundings. Each pixel is thus thermally isolated from its underlying support and circuitry with long, thin structures created by a thermal insulation cut for a large conductive thermal resistance R_t (Fig. 8.22).

Unfortunately, a long path length for thermal isolation may also result in low resonance frequencies; a folded thermal path can thus increase both the thermal resistance and resonance frequency (Fig. 8.23).

8.3.1.4 Interface Resistance

A type of thermal resistance to heat conduction depends on the roughness of the mating surfaces—the *interface* or *contact* resistance (Fig. 8.24). It results from air trapped in the gaps between rough surfaces, creating an excellent thermal insulator. Typical numbers to reduce the thermal resistance are a surface roughness of 32 microinches RMS or less. A very thin film of thermal "grease," indium foil, or other pliable material is also sometimes used between surfaces to fill in the air gaps with a material which is thermally more conductive than air. Care must be taken to not use too thick a film, however, as this then increases the thermal resistance via the increase in path length.

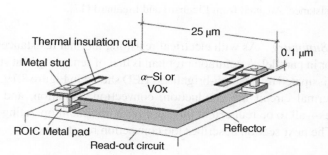

FIGURE 8.22 The long, thin structure created by the thermal insulation cut provides a large thermal resistance to minimize heat loss from the pixel to the surroundings. Adapted from E. Mottin et al., "Amorphous silicon technology improvement at CEA/LETI", Proc. SPIE, Vol. 4650 (2002).

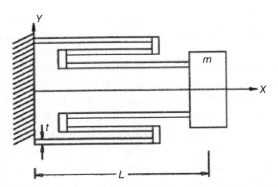

FIGURE 8.23 A folded structure can provide a thermal path length>L, while reducing the length that determines the resonant frequency of the structure. Credit: P. Thomas Blotter and J. Clair Batty, "Thermal and Mechanical Design of Cryogenic Cooling Systems," *The Infrared and Electro-Optical Systems Handbook*, Vol. 3, Chap. 6, SPIE Press (1993).

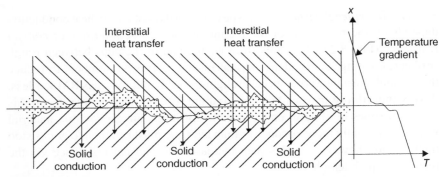

FIGURE 8.24 Air trapped in the interstitials between rough surfaces has a large thermal resistance. Adapted from Lienhard and Lienhard [17].

8.3.1.5 Summary As with electrical resistors, thermal resistances can add in series or in parallel. A common problem is that of removing heat such as the wattage dissipated by the high-brightness LED shown in Figure 8.19. Included in the thermal circuit are conduction, convection, radiation, and interface resistances—all to be reduced in this case for a minimum operating temperature T_{op}. The next section describes the convection resistance.

Example 8.2 Why does a focused beam heat up a material more than an unfocused beam? The power is the same in both cases, yet an image of the Sun on your hand with a so-called magnifying glass will easily burn your skin in microseconds, while the unfocused Sun will take a lot longer, and do much less damage. Why?

In the simplest model, the heat is transferred from the focused spot via conduction through an object—a disk or your hand, for example.[3] The surface area available for heat transfer depends on the size of the focused image or laser spot, with the smaller image having less area. With less area to transfer heat from the disk, the smaller image will have a higher temperature—see Equation 8.6 and Equation 8.7.

In practice, the heated surface itself will also transfer heat via the surrounding environment—air, water, or vacuum being the most typical. In the case of air or water, a heat-transfer mechanism known as *convection* may dominate over diffusion (Section 8.5.4), i.e., the radial conduction of heat away from the beam.

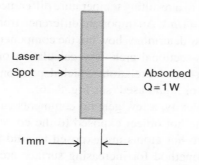

8.3.2   Convection

Example 8.2 ignores the convection from the top of the heated surface—that is, the flow of air or other fluids can increase the transfer of heat. The flow may be *natural* (due to the buoyancy of a heated fluid such as air—see Fig. 8.25) or *forced* (as with air fans or pumps for circulating water).

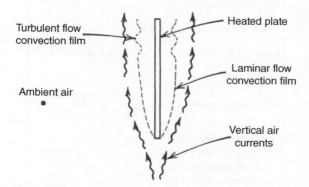

FIGURE 8.25 Hot air is less dense than cold, and rises against the less-energetic molecules, enhancing heat transfer through the process of natural convection. Credit: Dave S. Steinberg [15]

---

[3] Heat will also radially *diffuse* away from the heated surface—see **Section** 8.5.4.

Even though the basic heat-transfer equation for convection is similar to that for conduction, the calculation of the thermal resistance is extremely complex. The complexity arises in the calculation of the heat-transfer coefficient h (units of W/m^2-K), which depends on the air density, details of the fluid properties, the use of natural versus forced convection, and so on [17]:

$$Q = hA\,\Delta T \quad \text{[W]} \tag{8.8}$$

The equation for convective heat transfer thus hides the complexity of the physics, but is similar to Equation 8.6 for conductive heat transfer, in that a heat load Q is related to a resulting temperature difference ΔT via a convective thermal resistance $R_t = 1/hA$. An important difference from Equation 8.6 is that the surface area A now determines how hot the component will get, rather than the conductive cross-sectional area. The calculation of h is determined by highly empirical heat-transfer relations which depend on the fluid-mechanical properties of the gas or liquid used (see Fig. 8.26).

Despite the complexity, a few general comments can be made. First, the surface area A of the hot object exposed to the cooler fluid increases the heat removed—more hot atoms are exposed to cold atoms. For example, fins are a common method for increasing surface area. Fins that are "too

---

Convective heat transfer coefficient, h

• Reynolds number, Re $\qquad\qquad\qquad \text{Re} = \dfrac{\rho V L}{\mu}$

• Prandtl number, Pr $\qquad\qquad\qquad \text{Pr} = \dfrac{\mu c_p}{k}$

• Nusselt number, Nu $\qquad\qquad \text{Nu} = 0.23\,\text{Re}^{0.8}\,\text{Pr}^{0.4}$

• Convective coefficient, h $\qquad\qquad h = \dfrac{k\,\text{Nu}}{L}$

ρ = Fluid density (kg/m^3)
V = Fluid velocity (m/sec)
μ = Fluid viscosity (Pa)
c_p = Specific heat (J/g-K)

FIGURE 8.26 The calculation of the convective heat-transfer coefficient h has been empirically determined to depend on a fluid property known as the Nusselt number (Nu), which in turn depends on the Reynolds number (Re) and Prandtl number (Pr) in a complex way.

close"—depending on the height-to-spacing aspect ratio—can block (or "choke") convection. Fins also have heat-transfer efficiency, based on their length compared with the thickness (Fig 8.27). Fins that are thick have a higher efficiency, because the conduction area is larger, giving a smaller temperature drop along the length.

In general, forced convection is more efficient at transferring heat—that is, has a larger heat-transfer coefficient h—than natural convection. With forced convection, the supply of cold fluid is continuously being renewed (Fig. 8.28) at a faster rate. Alternatively, we can look at convection as increasing the surface area of the cold fluid exposed to the hot object, rather than the area of the hot object itself. For first-order analysis, the heat-transfer coefficients given in Figure 8.28 are used in Equation 8.8, rather than the complex formulation shown in Figure 8.26.

Figure 8.28 also shows that water is much more efficient at transferring heat than air. The reason is that water is denser, has a higher thermal conductivity, and a higher heat capacity—that is, an ability to absorb more heat for the same temperature rise—than air. Forced water cooling is significantly more complex

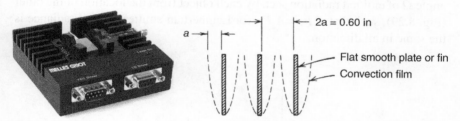

FIGURE 8.27 Fins can increase the surface area for convective heat transfer, and can be made efficient by controlling the height-to-width ratio. Photo credit: CVI Laser, LLC.

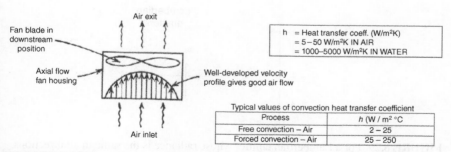

FIGURE 8.28 Forced-air cooling requires a fan, but increases the typical heat-transfer coefficient for air by a factor of approximately 10×. Dave S. Steinberg, *Cooling Techniques for Electronic Equipment* (2nd Edition), John Wiley & Sons (1991).

than air cooling, however—requiring pumps, expansion tanks, and leak-free fittings—and should only be used when air cooling is clearly insufficient.

8.3.3 Radiation

We mentioned in Example 8.2 that heat can also be transferred across an evacuated space (i.e., vacuum). In this case, electromagnetic energy is radiated (emitted) by *everything* at a rate proportional to the 4th-power of its temperature. Objects at different temperatures thus transfer heat from hot to cold at a net rate Q_{12} that depends on their temperature difference and the surface area A_1 of the hot object [17]

$$Q_{12} = \varepsilon\, \sigma\, A_1\, F_{12}(T_1^4 - T_2^4) \quad [\text{W}] \tag{8.9}$$

For room temperature (RT) surroundings, radiation is an important heat-transfer mechanism for objects at temperatures greater than approximately 400 K, or in a vacuum where convection is low. An additional factor F_{ij}—known as the *view factor*—has also been included to account for the solid angle Ω of emitted radiation seen by each object from the location of the other (Fig. 8.29), such that $A_1 F_{12} = A_2 F_{21}$ for Lambertian emitters whose radiance is the same in all directions.

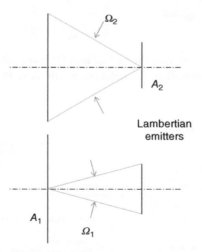

FIGURE 8.29 For a Lambertian emitter whose radiance is the same in all directions, the view factor F_{12} is the fraction of power emitted from Surface 1 intercepted by Surface 2, as determined by the solid angle Ω_1. A view factor F_{21} can also be defined, such that $A_1 F_{12} = A_2 F_{21}$.

The heat transferred also depends on the effective emissivity ε of the objects (with a range of values from 0 to 1). For opaque objects, the emissivity is a *surface* property, with "shiny" objects (high reflectivity at wavelength λ) emitting very little [$\varepsilon(\lambda)$ small, <0.1], and "dark" objects (low reflectivity at wavelength λ) emitting a great deal [$\varepsilon(\lambda)$ large, >0.9]. Objects which absorb everything do not transmit and do not reflect. If their temperature is not changing, energy conservation requires that they must emit whatever energy is absorbed. This is *Kirchhoff's Law*: absorption $\alpha(\lambda)$=emission $\varepsilon(\lambda)$, so good absorbers at a particular wavelength λ are also good emitters at the same wavelength, when the objects are not heating up or cooling down (i.e., are in thermal equilibrium).

A low emissivity is used when insulating against heat transfer (such as keeping the optics in a cold dewar at a low temperature). A high emissivity is used when augmenting heat transfer (such as intentionally cooling an instrument via radiation). The wavelength dependence of $\varepsilon(\lambda)$ indicates that what is emissive at one wavelength may be reflective at another. This can be used to our advantage, as a surface with a low solar absorption α_s and high long-wave IR emissivity ε_{IR}—a typical white paint, for example—is excellent for passive cooling of room temperature objects which emit at a peak wavelength $\lambda \approx 10\,\mu m$.

Various surface coatings, films, and paints are available to engineer the emissivity of a surface. Some of these coatings use engineered materials to increase the absorption area of the film on a microscopic scale by increasing the number of bounces incident light sees (Fig. 8.30a). Increasing the absorption area thus increases the effective emissivity. Similarly, decreasing the absorption area—by superpolishing a mirror to a very smooth (low) surface roughness—has the opposite effect, and can be used to reduce the effective emissivity.

The same effect can be seen on a macroscopic scale, where convection fins can also have a significantly higher emissivity than implied by their surface emissivity alone. Each fin acts like a cavity, with the increase in emissivity depending on the depth L of the fin compared with its spacing w from the other fins (Fig. 8.30b).

When the emissivities of two objects are not the same, the net heat transfer Q_{12} depends on the emissivities and the geometry of both. An effective emissivity is then used in Equation 8.9 for radiative heat-transfer calculations, where Figure 8.31 shows that the reciprocal effective emissivity $1/\varepsilon = 1/\varepsilon_1 + 1/\varepsilon_2 - 1$ for large parallel planes, for example. The figure also illustrates the view factor $F_{12} = 1$ for the different geometries.

The general equation describing the net heat transfer between surfaces with different ε is given by [17]

$$Q_{12} = \frac{\sigma A (T_1^4 - T_2^4)}{\dfrac{1-\varepsilon_1}{\varepsilon_1 A_1} + \dfrac{1}{A_1 F_{12}} + \dfrac{1-\varepsilon_2}{\varepsilon_2 A_2}} \quad [W] \qquad (8.10)$$

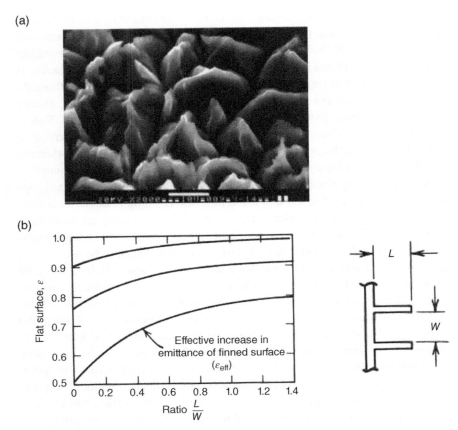

FIGURE 8.30 (a) A rough surface on a microscopic scale has a large absorption, and therefore large emissivity; (b) Convection fins illustrate the same physics, where the effective emissivity of the fins is larger than that of each surface individually. Credits: (a) Reprinted with permission from M. J. Persky [18], Copyright 1999, AIP Publishing LLC; (b) Dave S. Steinberg *Cooling Techniques for Electronic Equipment* (2nd Edition), John Wiley (1991).

where the view factor F_{12} is now included. This factor is the fraction of power emitted by surface 1 that is intercepted by surface 2. If the surfaces are diffuse Lambertian radiators emitting into 2π steradians—that is, they have the same radiance in all directions—then $A_1F_{12} = A_2F_{21}$, as shown in Figure 8.29. Many geometries are more complex than those shown in Figure 8.31, and determining the view factor requires the use of design curves which are available in heat-transfer textbooks—see Refs. [17] and [19] for more details.

A typical example of radiation heat transfer is the thermal design of vacuum dewars for cooled (80 K) FPAs. The vacuum, with approximately

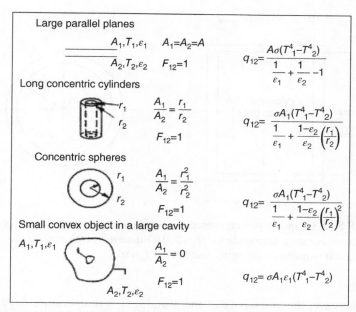

FIGURE 8.31 Geometric dependence of the effective emissivity used in Eq. (8.9). Credit: P. Thomas Blotter and J. Clair Batty, "Thermal and Mechanical Design of Cryogenic Cooling Systems," *The Infrared and Electro-Optical Systems Handbook*, Vol. 3, Chap. 6, Fig. 6.21, SPIE Press (1993).

zero thermal conductivity, insulates the FPA from the 300 K temperature of the dewar structure. However, radiation across the dewar vacuum is a heat load on the cryogenic cooler required to maintain the FPA at 80 K. The lower this radiation load, the smaller the cooler needs to be, and the greater the size, weight, and cost savings.

Example 8.3 An aluminum tank containing liquid nitrogen (LN_2) maintains an 80 K temperature for the FPA detector inside a dewar (Fig. 8.32), and has a finite lifetime based on the heat transferred from the surroundings at $T_2 = 300$ K. How much heat is transferred from the surroundings to the optics and the FPA?

The geometry shown in Figure 8.32 is that of two concentric cylinders. From Figure 8.31, the heat transfer between concentric cylinders with $F_{12} = 1$ is given by:

$$Q_{21} = \frac{\sigma A_1 (T_2^4 - T_1^4)}{\dfrac{1}{\varepsilon_1} + \dfrac{1-\varepsilon_2}{\varepsilon_2}\dfrac{R_1}{R_2}}$$

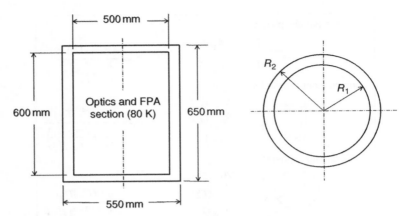

FIGURE 8.32 Geometry of a cryogenic dewar, where conduction and radiation from the room temperature surroundings ($T_2 = 300\,\text{K}$) determines in part the size of the cooler needed to maintain the optics and FPA at $T_1 = 80\,\text{K}$.

In this case, the emissivity of the aluminum is temperature dependent, with $\varepsilon_2 = 0.08$ at 300 K and $\varepsilon_1 = 0.04$ at 80 K [7]. In addition, the radiative surface area $A_1 = \pi DL = \pi(0.5\,\text{m})(0.6\,\text{m}) = 0.94\,\text{m}^2$. Substituting in the above equation

$$Q_{21} = \frac{5.67E-8\,\text{W/m}^2-\text{K}^4 \times 0.94\,\text{m}^2 \times \left[(300K)^4 - (80K)^4\right]}{\dfrac{1}{0.04} + \dfrac{1-0.08}{0.08}\left(\dfrac{0.5\,\text{m}}{0.55\,\text{m}}\right)} = 12.1\,\text{W}$$

There is an additional heat load on the LN_2 tank from the end plates of approximately 5 W, giving a total heat load of approximately 17 W. If the cooler removing this heat from the LN_2 tank is only 5% efficient at converting electrical power into cooling power, it will require on the order of 340 W of electrical power—a huge number for any space-based optical system.

For this system, the tank will instead be filled with cryogen, whose lifetime in space is determined by the 17 W of radiative heating, as well as any conductive heat transfer and the power consumption by the FPA. Radiation shields and multi-layer insulation (MLI) can be used between the cylinders to reduce the heat transferred, and extend the life of the nitrogen supply. The same concept is sometimes used to protect agricultural crops from freezing on cold nights, where "smudge" pots provide a smoke barrier against long-wave IR radiative heat loss from the crops to the cold night sky.

8.4 THERMAL MANAGEMENT

The ability to engineer the emissivity of surfaces to control temperatures and heat transfer introduces an important topic known as *thermal management*. That is, given the various heat sources mentioned in Section 8.3—heaters, electronics, solar loading, friction (mechanisms), ambient temperature of 300 K (for cooled systems), dissipated power due to the wall-plug inefficiency of sources (lasers, LEDs, etc.), and so on—we are now in a position where we can estimate how hot the system gets using the principles of conduction, convection, and radiation.

Thermal management extends this knowledge to the engineering level by looking at the thermal technologies available for heat generation, heat removal, and control of temperature distributions. For example, heat generation is not necessarily something to be avoided, and heaters are commonly engineered into consumer telescopes to prevent condensation on the optics during night viewing. Similarly, reducing detector temperature for a high-sensitivity camera—a high-end CCD array for biomedical imaging, for example—reduces the inherent noise of the CCD [20]; in this case, a thermoelectric cooler (TEC) is used to reach temperatures as low as minus 55 C in still air—a much less expensive technology than cryogenic liquids.

In general, the thermal management technologies available to control absolute temperature as well as thermal gradients include the following:

- Heaters—to maintain a constant laser or telescope temperature, for example
- Heat spreaders—such as thin diamond films to spread heat over a large area
- Air cooling fins—natural (passive) and forced convection
- Air cooling fans—forced convection for moderate heat loads
- Liquid cooling pumps—forced convection for high heat loads
- Thermoelectric (solid-state) coolers—for temperatures $T_{op} > 200$ K
- Heat pipe (evaporative) coolers—for low ΔT with high heat loads
- Stirling-cycle coolers—for cryogenic temperatures $T_{op} < 150$ K
- Thermal "greases" and thermal interface materials—control contact resistance
- Coatings and paints—control reflectivity and ratio of absorption to emissivity (α/ε)
- Materials selection—control absorption, CTE, thermal conductivity, etc.

In the following sections, we look in more detail at the use and specification of four of these components: heaters, fans, thermal interface materials, and thermoelectric coolers.

8.4.1 Heaters

While many different types of heaters are available, the resistive heaters embedded in flexible Kapton polyimide are commonly used for optical instruments. Consumer and space-based telescopes both use these heaters to control temperature and temperature profile—to keep the mirrors hot enough to prevent condensation (both consumer and space-based), and to keep temperature distributions from creating excessive misalignments (space-based).

The power density (W/cm² or W/in²) of such heaters is limited by the melting of Kapton in which the heaters are embedded. Figure 8.33 also shows that above a certain temperature, the power density is limited by the temperature of the environment; this is expected, as the hotter environment will result in a higher Kapton temperature for a given heat load. The Kapton is also a low-outgassing material, thus allowing their use for space-based telescopes where volatiles can condense on cold optical surfaces, reducing the transmitted photons and increasing unwanted background scatter.

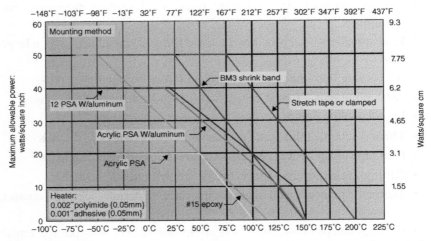

FIGURE 8.33 Resistive heaters embedded in Kapton are commonly used for controlling the temperature of telescopes. Adapted from Minco (www.minco.com).

8.4.2 Fans

Small "muffin" fans are common thermal management components for optical systems. Such fans can supply a certain flow rate of air (liters/sec or CFM = ft³/min) for convective cooling, while pushing against a certain static back-pressure (Pascals) from the acoustic impedance of the volume being cooled (i.e., a telescope tube). They are typically specified by the vendors for the simple situation where they are used in isolation from their application. Figure 8.34, for example, shows that fan "A" is specified as a "120 CFM" fan; the performance plot for Fan A shows that this is the air flow that can be supplied only if there is zero back pressure.

As just about any volume that needs to be cooled will have some degree of back pressure, the maximum value of 120 CFM is not a realistic air flow that can be expected for fan "A." Instead, the fan will supply some lower flow rate—thus reducing the amount of heat that can be transferred for a required temperature change. The actual flow rate will depend on how the back pressure of the cooled volume increases with flow rate. This is also shown in Figure 8.34, where the curves labeled D, E, and F correspond to nominal, good (lower), and excellent (even lower) volume impedance. The intersection of each of these curves with a given fan type (A, B, or C) determines the air flow that can be expected, and thus the cooling capacity of forced convection—see Equation 8.8 and Figure 8.26.

8.4.3 Thermal Interface Materials (TIMs)

Thermally conducting electrical insulators are commonly used to remove heat from electronic components without electrical shorting. TIMs are flexible,

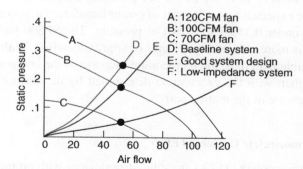

FIGURE 8.34 The intersection of the fan's impedance curve with that of the cooled volume determines the fan's flow rate and resulting cooling capacity. Adapted from Comair Rotron.

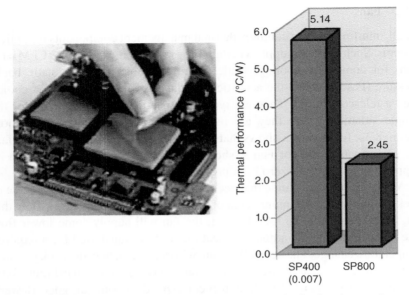

FIGURE 8.35 Thermal interface materials are used to fill in the air gaps between materials with poor surface finish, thus reducing the thermal resistance. Reproduced by permission of the Bergquist Company.

allowing efficient heat transfer across surfaces that are not flat, smooth, or have other irregularities. By filling in the low-conductivity air gaps, TIMS reduce thermal contact resistance. Two common products are Chomeric's TIM and Bergquist's Sil-Pad (Fig. 8.35).

The performance of a TIM is measured in terms of its thermal resistance, measured in units of K/W (or C/W). For example, a temperature increase of 5.14 C can be expected for every watt of power transferred across a Bergquist SP400 TIM under 0.34 MPa (50 psi) of pressure. The SP800 part shown in Figure 8.35 is more efficient by a factor of approximately 2. Without the use of these flexible materials, the temperature increase of the component would be much higher, with the exact value determined by the surface finish and clamping pressure of the mating parts.

8.4.4 Thermoelectric Coolers (TECs)

Thermoelectric coolers (TECs) are solid-state devices with no moving parts, and create cold surfaces with electric current at the junction of dissimilar materials (Peltier effect—see Fig. 8.36). At this time, they are fairly *in*efficient at converting electrical power into cooling capacity, so the heat created—and

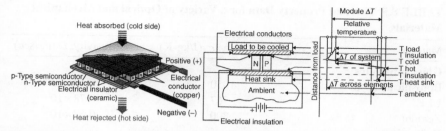

FIGURE 8.36 Thermoelectric coolers are solid-state devices with no moving parts, using the Peltier effect to reduce the temperature of a junction with electrical current. Adapted from Melcor Corp.

that must also be removed—is large compared with the number of watts of cooling. Their efficiency also depends on the temperature difference that is required, with a larger ΔT being less efficient.

As with fans, TECs are specified by the vendors for the situation where they are used in isolation from their application. For fans, it was the air flow that can be supplied for zero back pressure; for TECs, it is the highest temperature difference that can be created for zero heat load. This is, of course, not a realistic design situation, as the TEC is specifically intended to remove a finite load. These loads can be added to the calculations, and the vendors often have software to facilitate this task, but the spec sheet performance data will not generally reflect this necessary level of detail.

8.5 MATERIAL PROPERTIES AND SELECTION

Materials selection is also an important part of thermal management. To keep components from overheating, for example, a large thermal conductivity is required. Alternatively, to keep components such as FPAs at a low temperature, a low thermal conductivity is required between the FPA and the warmer world. The materials-selection parameters expanded on in this section include the following:

- Thermal expansion
- Thermal distortion
- Thermal mass
- Thermal diffusivity
- Thermal shock

TABLE 8.5 Thermal Property Data for a Variety of Optical and Mechanical Materials

Material	Heat capacity, c_p (J/kg-K)	Diffusivity, D (m²/sec)
Glass substrates		
Borosilicate (Pyrex)	1050	0.48×10^{-6}
ULE	767	0.78×10^{-6}
Zerodur	800	0.67×10^{-6}
Refractives		
Fused silica (SiO_2)	750	0.79×10^{-6}
Germanium (Ge)	310	36×10^{-6}
N-BK7	860	0.52×10^{-6}
Silicon (Si)	753	93×10^{-6}
Zinc selenide (ZnSe)	340	10×10^{-6}
Zinc sulfide (ZnS)	500	13×10^{-6}
Structural		
Aluminum (6061-T6)	900	69×10^{-6}
Beryllium (I-70H)	1930	60×10^{-6}
Copper (OFHC)	385	114×10^{-6}
Graphite epoxy	920	20×10^{-6}
Invar 36	510	2.53×10^{-6}
Silicon carbide (Si-SiC)	140	411×10^{-6}
Stainless steel (304)	500	4.12×10^{-6}
Titanium (6Al-4V)	530	2.88×10^{-6}

Many of these properties are given in Table 8.1 and Table 8.5 for a variety of optical and mechanical materials. Ultimately, the choice of materials will involve a balance of competing requirements—and these requirements will determine not only the acceptable thermal parameters but also the structural and vibration parameters reviewed in the previous chapters as well.

8.5.1 Thermal Expansion

We saw in Section 8.1.1 that the CTE depends on temperature. However, the CTE α_t of some materials—composites, for example—can be controlled by the vendors to very low values at a particular temperature. The CTE of ULE and Zerodur, for example, can be tailored via materials composition to be extremely low—and exactly zero at a specific temperature (Fig. 8.37). Graphite-epoxy (GrE) composites can be similarly fabricated, allowing the near-zero expansion of the metering structure maintaining the Hubble Space Telescope's 5-meter primary-to-secondary mirror spacing.

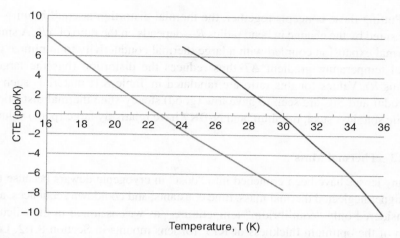

FIGURE 8.37 The CTE of some materials can be tailored to be exactly zero at a specific temperature. Data from SCHOTT Technical Note TIE-37, "Thermal expansion of Zerodur."

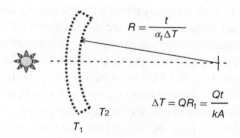

FIGURE 8.38 The distortion [or curvature change $\Delta C = \Delta(1/R)$] of a material absorbing a heat load Q depends on the CTE α_t and the thermal conductivity k, such that $R \approx kA/\alpha_t Q$ for an unconstrained mirror which is initially flat.

8.5.2 Thermal Distortion

As we have seen, a material exposed to a heat source such as the Sun will bend and distort. This occurs when the side exposed to the source thermally expands more than the shaded side—a result of a temperature difference created by the thermal resistance of the material. As shown in Figure 8.38, the distortion is bigger for a larger CTE α_t—that is, the radius R becomes smaller (less planar) as one side of the material expands more than the other due to the temperature gradient ($\Delta T = T_1 - T_2$). At the same time, the steady-state axial temperature gradient depends on the thermal resistance of the material, such that a low-conductivity material results in a larger ΔT for a given heat absorption Q.

Putting these concepts together, the thermal distortion at equilibrium—as measured by the change in bend radius R—depends on the ratio of α/k. A small thermal expansion coupled with a large thermal conductivity to minimize the axial temperature gradient ΔT thus reduces the distortion (has the largest radius R). Values for this ratio are tabulated in Table 8.1; materials such as Zerodur and SiC are seen to have low (good) steady-state thermal distortion, while stainless steel, titanium, and N-BK7 have relatively large (poor) values.

8.5.3 Thermal Mass

Many lenses have been crunched into "dust" in cryogenic dewars because the designer neglected thermal mass, time constants, and cooldown sequences, and considered only "equilibrium" temperature—as was assumed in the calculation of the optimum thickness of RTV for lens mounts in Section 8.1.2. Less catastrophically, the large mass of ground-based astronomy telescopes prevents them from rapidly changing temperature, thus not only preventing changes in figure for short-term changes in ambient temperature, but also resulting in lost observation time as longer-scale temperature changes—on the order of a week, for example—slowly work their way through the telescope's massive mirrors.

The critical factor in these cases where time is now a variable is the thermal mass and thermal time constant of the optics. Thermal mass m_t is the actual mass m weighted by a thermal parameter known as the specific heat c_p, measured in units of J/kg-K.[4] That is, $m_t = mc_p = \rho V c_p$, with units of joules per degree Kelvin (J/K). So thermal mass—also known as thermal capacitance or thermal inertia—is a measure of the energy change for a given temperature change. It determines, in part, the time it takes for a mirror, telescope, or entire optical system to heat up or cool down.

Depending as it does on the density and specific heat, thermal mass can be reduced with appropriate material selection (see Problem 8.11). Since it also depends on the volume V of material used, it can also be reduced using the structural lightening features reviewed in Section 5.4, while still maintaining structural rigidity.

However, the thermal time constant depends on more than the thermal mass; we must also include how well the optics are thermally connected to the heat source (or sink) that is changing its temperature—that is, the thermal resistance to conduction, convection, and radiation. Just as with electrical circuits, whose time constant depends on both the resistance R and capacitance C of the circuit, the thermal time constant also depends on the thermal resistance

---

[4] The specific heat c_p is also known as specific heat capacity or simply the heat capacity. The subscript "p" indicates that it is measured at a constant pressure, which for incompressible materials such as aluminum and fused silica is identical to a measurement at a constant volume (giving $c_p = c_v$). See Ref. [17] for more details.

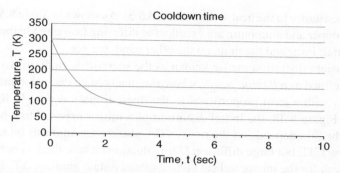

FIGURE 8.39 The equilibrium temperature of an optical component—in this case, 80 K—is obtained at approximately 5τ (for $\tau = 1$ second in this example).

R_t. Thermal mass and thermal resistance thus both determine how long an optical system will take to cool down or heat up, giving a thermal time constant $\tau = R_t C = R_t m_t$. An example of how the temperature of an optical component changes with time is shown in Figure 8.39; the figure illustrates that approximately 5τ is required for the component to reach thermal equilibrium.

8.5.4 Thermal Diffusivity

While the thermal time constant indicates how long it takes for a lens to heat up or cool down based on its thermal mass and connection to the environment, a related concept is how long it takes for heat to spread out from a source *within* a material. For example, a laser beam focused onto a metal has little surface area to dissipate the absorbed heat, and the immediate temperature rise at focus could depend more on the material's inherent ability to spread heat, and less on its thermal time constant with the outside world via conduction, convection, and radiation.

The spreading out (or *diffusion*) of heat is measured as a temperature change, and the time for the change to occur depends on the distance from the source (units of meters) and a material property known as the thermal diffusivity D (units of m²/sec). The thermal diffusivity depends in turn on the thermal conductivity k of the material, where a higher conductivity diffuses heat more readily. In addition, the thermal diffusivity also depends inversely on the mass density ρ—but not the mass m—such that a higher density prevents the diffusion of heat, leading to slower temperature changes. Quantitatively, $D = k/\rho c_p$, where a large specific heat c_p also prevents the heat from spreading rapidly from the source; the resulting diffusion time constant $\tau_d = L^2/D$, where L is the distance from the heat source. Depending as it does on material properties (k, ρ, and c_p), this time constant may be looked at as a material's inherent RC time constant—as distinct from a time constant which depends on the thermal resistance to the heat sink (Section 8.5.3). As shown in Table 8.5, metals such as copper and aluminum are excellent at diffusing heat quickly throughout the material; invar and titanium, on the other hand, are not.

A transient distortion metric known as the specific thermal diffusivity can also be defined as $\alpha_t/D = \rho c_p \times \alpha_t/k$, where the steady-state distortion metric α_t/k combines with a term proportional to the thermal mass $m_t = \rho c_p V$. Referring again to Figure 8.38, the initial distortion of a mirror exposed to a heat load such as the Sun depends on the ratio α_t/D—placing an emphasis on materials with a low CTE but large diffusion D for situations where there is not enough time to wait for the mirror to heat up to its steady-state gradient ΔT. Table 8.1 shows that aluminum and copper are reasonable materials for such applications—assuming that the steady-state distortion is also acceptable.

8.5.5 Thermal Shock

Thermal transients that occur too quickly can stress a material beyond its strength. For example, quenching a hot ceramic in cold water can break the ceramic—as many cooks discovered over the years before CorningWare was invented in 1958. A material parameter S' is used for evaluating the resistance to such thermal shock [21]

$$S' = \frac{\sigma_f k (1 - v)}{\alpha_t E} \tag{8.11}$$

A low thermal distortion α_t/k gives better shock resistance (large S'), as does a large material strength σ_f and a *small* modulus E for flexibility. This thermal parameter must be used with caution, as the fracture strength σ_f for brittle materials depends strongly on the flaw sizes present on the surface of

TABLE 8.6 The Thermal Shock Parameter S' Depends on the Ratio $k/\alpha_t E$[a]

Material	CTE, α_t (ppm/K)	Elastic modulus, E (MPa)	Thermal conductivity, k (W/m-K)	Poisson's ratio, μ	Fracture strength, σ_f (MPa)	Thermal shock, S' (W/m)
LiF	37	85	14.2	0.27	11	37
BaF2	17	65	7.1	0.31	27	120
ZnSe	7.6	70	16	0.28	50	1083
Fused Quartz	0.6	73	1.4	0.17	60	1611
ZnS	7.0	74	19	0.29	100	2604
Sapphire	5.3	344	36	0.27	300	4324
Germanium	6.1	103	59	0.28	90	6085

[a]Data from Hasselman [21]

the glass—see Section 6.2. As shown in Table 8.6, germanium and sapphire have good resistance to thermal shock, while LiF_2 and BaF_2 do not for the strengths listed—and these data can easily be scaled for other strengths. In practice, fused quartz does not have the highest shock resistance, but is commonly used in high-power lamps as a low-cost material with good optical transmission and thermal-shock protection.

PROBLEMS

8.1 How do the rabbit's ears change to help the rabbit stay warm in the winter?

8.2 The Hubble Space Telescope uses graphite-epoxy (GrE) metering tubes to maintain a primary-to-secondary mirror despace of less than $3\,\mu m$ over a 5-m distance. The CTE of these tubes can be tuned by varying the fraction of epoxy and by changing orientation of the graphite fibers with respect to the optical axis. What CTE is required for a uniform on-orbit temperature change of $\Delta T = 30\,K$? What assumptions are being made with respect to the CTE that could modify this calculation?

8.3 Using Figure 8.11, show that the unconstrained thermal expansion of a lens gives the change in focal length $\Delta f = \alpha_t f \Delta T$. Hint: the change in radii changes the power of the lens—see Chapter 2.

8.4 A long-wave IR imager uses a thin f/2 germanium lens with a 100-mm focal length in an aluminum mount. What temperature change is allowed before the image quality is unacceptable? Does the focus move toward or away from the focal plane as the temperature increases?

8.5 The designer of a high-power laser system is selecting a window material. Compare and contrast the properties of fused silica, sapphire, and N-BK7. Are there other properties besides thermal distortion which must be taken into account?

8.6 A high-brightness LED is overheating, and it is proposed that a 1-mm thick spacer of copper—selected for its high thermal conductivity—should be added between the LED and the ceramic substrate. Using the 1D heat-transfer model shown in Figure 8.19, what is the temperature change with and without the copper spacer? Note that: (1) the LED dissipates 1 W of heat; (2) without the copper spacer, the total thermal resistance from the LED to the heat sink at T_s is $R_t = 10\,K/W$; and (3) the spacer is the same size as the LED (1 mm × 1 mm). What is the optimum thickness for such a spacer? Is there a way to modify the spacer to improve the temperature change?

8.7 In Section 8.3.3, it was stated that a good absorber is a good emitter. But what does this really mean? Specifically, a surface that is black to our eyes is clearly a good absorber—so why can't we also see that the object is emitting light?

8.8 In Figure 8.34, what flow rate is expected for a 120 CFM fan with the nominal, good, and excellent volume impedance? What about a 70 CFM fan?

8.9 Physically, why doesn't the thermal distortion of a mirror (Fig. 8.38) depend on the thickness t of the mirror?

8.10 Referring to Figure 8.38, estimate the thermal deformation of a window exposed to the Sun on one side. Hint: see Ref. [22].

8.11 Using Table 8.5, add a column listing the product of the density ρ and specific heat c_p. Which of the materials has the lowest thermal mass?

8.12 A designer is using a very thin, lightweight membrane (pellicle) as an optical filter for a ground-based telescope which will survey the cold night sky. He is having a difficult time finding a pellicle material which matches the CTE of its mount. Can you help? Hint: what is the thermal mass of the membrane compared with the mount? Is a CTE match even necessary?

8.13 How does CorningWare prevent thermal shock from breaking hot dinner plates that have been quenched in cold water?

8.14 In selecting refractive materials for thermal design, indicate whether the following parameters should be large or small: absorption coefficient α_m, CTE α_t, thermal slope dn/dT, thermal conductivity k, and the steady-state distortion ratio α_t/k. Do your answers depend on the application? If so, give examples of when it is better for a parameter to be large in one application yet small in another.

REFERENCES

1. R. E. Fischer, B. Tadic-Galeb, and P. R. Yoder, *Optical System Design* (2nd Edition), New York: McGraw-Hill (2008), Chaps. 16–18.

2. R. A. Paquin, "Properties of metals," in M. Bass, E. W. Van Stryland, D. R. Williams, and W. L. Wolfe (Eds.), *Handbook of Optics* (2nd Edition), Vol. 2, New York: McGraw-Hill (www.mcgraw-hill.com) (1995), Chap. 35.

3. M. J. Weber, *Handbook of Optical Materials*, Boca Raton: CRC Press (www.crcpress.com) (2003).

4. *The Crystran Handbook of Infra-Red and Ultra-Violet Optical Materials*, Crystran Ltd (www.crystran.co.uk) (2008).

5. P. Giesen and E. Folgering, "Design guidelines for thermal stability in opto-mechanical instruments," Proc. SPIE, Vol. 5176, pp. 126–134 (2003).

6. K. B. Doyle, V. L. Genberg, and G. J. Michels, *Integrated Optomechanical Analysis* (2nd Edition), Bellingham: SPIE Press (2012).

7. P. T. Blotter and J. C. Batty, "Thermal and mechanical design of cryogenic cooling systems," in W. D. Rogatto (Ed.), *The Infrared and Electro-Optical Systems Handbook*, Vol. 3, Bellingham: SPIE Press, pp. 343–433 (1993), Chap. 6.

8. D. Vukobratovich, *Introduction to Optomechanical Design*, SPIE Short Course SC014 (www.spie.org) (2009).

9. M. Bayer, "Lens barrel optomechanical design principles," Opt. Eng., Vol. 20, No. 2, pp. 181–186 (1981).

10. P. J. Rogers and M. Roberts, "Thermal compensation techniques," in M. Bass, E. W. Van Stryland, D. R. Williams, and W. L. Wolfe (Eds.), *Handbook of Optics* (2nd Edition), Vol. 1, New York: McGraw-Hill (1995), Chap. 39.

11. M. Bass, E. W. Van Stryland, D. R. Williams, and W. L. Wolfe (Eds.), *Handbook of Optics* (2nd Edition), Vol. 1, New York: McGraw-Hill (www.mcgraw-hill.com) (1995), Chap. 39, Table 3.

12. W. J. Smith, *Modern Optical Engineering* (4th Edition), New York: McGraw-Hill (2008), Chaps. 10, 14, 16, 20.

13. N. W. Wallace and M. A. Kahan, "First-order thermo-optical and optomechanical wavefront error analysis," Proc. SPIE, Vol. 3030, pp. 109–120 (1997).

14. H. Weichel, "Lasers," in W. D. Rogatto (Ed.), *The Infrared and Electro-Optical Systems Handbook*, Vol. 3, Bellingham: SPIE Press, pp. 575–650 (1993), Chap. 10.

15. D. S. Steinberg, *Cooling Techniques for Electronic Equipment* (2nd Edition), New York: John Wiley & Sons (1991).

16. E. Mottin, A. Bain, P. Castelein, J.-L. Ouvrier-Buffet, J.-L. Tissot, J.-J. Yon, and J.-P. Chatard, "Amorphous silicon technology improvement at CEA/LETI," Proc. SPIE, Vol. 4650, pp. 138–149 (2002).

17. J. H. Lienhard, IV and J. H. Lienhard, V, *A Heat Transfer Textbook* (3rd Edition) Cambridge, MA: Phlogiston Press (2008).

18. M. J. Persky, "Review of black surfaces for space-borne infrared systems," Rev. Sci. Instrum., Vol. 70, No. 5, pp. 2193–2217 (1999).

19. W. Rohsenow, J. Hartnett, and Y. Cho, *Handbook of Heat Transfer* (3rd Edition), New York: McGraw-Hill (www.mcgraw-hill.com) (1998).

20. K. J. Kasunic, *Optical Systems Engineering*, New York: McGraw-Hill (www.mcgraw-hill.com) (2011).

21. D. P. H. Hasselman, "Figures-of-merit for the thermal stress resistance of high-temperature brittle materials: a review," Ceramurg. Int., Vol. 4, pp. 147–150 (1978).

22. W. P. Barnes, Jr., "Some effects of aerospace thermal environments on high-acuity optical systems," Appl. Opt., Vol. 5, No. 5, pp. 701–711 (1966).

9

KINEMATIC DESIGN

The tripod is a common method for mounting an optical system, but it has an unfortunate disadvantage: it can be easily moved when accidentally "bumped," a typical occurrence during photography sessions. The motions that lead to changes in alignment—and loss of long-exposure pictures—are linear (sliding along the ground) and rotation (about any axis perpendicular to the ground). Such kinematic instabilities are due to the fact that the tripod's three legs provide only three constraints on its motion, with three degrees of freedom (DOFs) remaining: two linear and one rotational. One goal of this chapter is to answer the question: What would it take to constrain the tripod in all six DOFs, in a way that is insensitive to environmental effects such as temperature changes?

As shown in Figure 9.1, any optic has six DOFs—with three coordinate axes, there are three linear motions along the x-, y-, and z-axes, as well as three rotations about the axes. A strain-free, kinematic mechanism will constrain these six motions only once. An optical element that is "overconstrained"— that is, holds on to the optic or its mount in ways that constrain any of the DOFs *more than once*—strains it and the optic, inducing WFE.

The concepts of optical DOFs such as tilt, decenter, and despace will be explored in Section 9.1. We then apply these ideas to the mounting of optical components (Section 9.2), as well as review in Section 9.3 the alignment

Optomechanical Systems Engineering, First Edition. Keith J. Kasunic.
© 2015 John Wiley & Sons, Inc. Published 2015 by John Wiley & Sons, Inc.

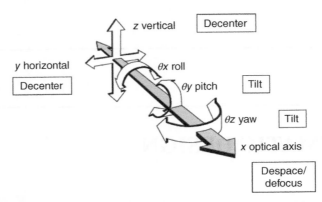

FIGURE 9.1 Any optic has six degrees of freedom. Credit: CVI Laser, LLC.

mechanisms that may be needed if a DOF (or two) needs to be adjusted during assembly or use. Finally, Section 9.4 looks in more detail at the key material properties for mounts and mechanisms.

9.1   KINEMATIC AND SEMI-KINEMATIC MOUNTS

A *kinematic mount* is one which constrains all six possible motions (DOFs) only once using ideal "point" contacts. The motivations for using kinematic mounts are:

- Accurate, repeatable positioning (both position and angle) of an optical component with respect to a reference location;
- Minimum strain on the mount and the optics for static loads, vibrations, or a change in temperature. As we will see, constraining a DOF more than once results in excess strain and WFE that can be minimized with the use of a kinematic mount.

One approach to a kinematic mount is shown in Figure 9.2. The figure shows the use of an arrangement—credited to Lord Kelvin—known as a "vee–cone–flat" to constrain all six DOFs. For this design, the cone has line contact with the ball which constrains linear motion along x-, y-, and z-axes, the vee provides two-point contact that constrains rotation about y- and z-axes (pitch and yaw in Fig. 9.2a), and the flat has one-point contact which constrains rotation about the roll axis in Figure 9.2a. The ideal line contact of the cone is not strictly kinematic, and can be replaced with a trihedral contact to make it so.

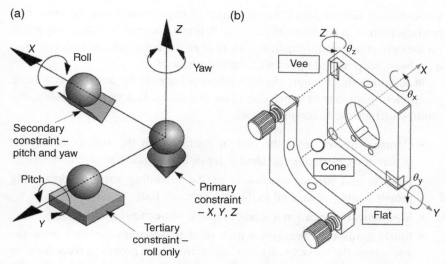

FIGURE 9.2 The kinematic constraint of six degrees of freedom is illustrated in (a); the hardware implementation is shown in (b). Credits: (a) Permission to use granted by Newport Corporation; all rights reserved; (b) CVI Laser, LLC.

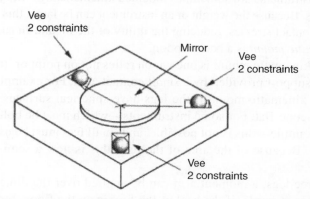

FIGURE 9.3 A three-vee kinematic mount does not result in decenter when the temperature of the mount and the optic change uniformly. Credit: Optical Society of America [1].

The vee–cone–flat design shown in Figure 9.2 is not a unique way to kinematically constrain all six DOFs. Another design is shown in Figure 9.3, where three vees are arranged in a radial pattern to obtain a kinematic mount which is more stable during temperature changes [2]. Each vee constrains two DOFs—one with each side of the vee—and the vees all point to the center of the optical component, preventing the optic from moving if the vees uniformly change position due to thermal expansion or contraction. The more traditional vee-cone-flat Kelvin mount has a center of expansion about the cone that restricts motion in the plane of the optic. While it does not introduce any strain on the optic, the thermal expansion away from the cone introduces decenter—a problem for powered optics with curved surfaces.

In addition to temperature changes, practical issues in the design of kinematic positioners—any one of which can cause instability, lack of repeatability, and strain—include the following:

- Contact stresses from the ball in contact with the vee, cone, or flat (Chapter 6), changing the ideal point or line contacts to area contacts;
- Friction and surface roughness of the mounting surfaces preventing proper seating of the ball in the vee, cone, or flat;
- Mechanical fabrication tolerances of the vee-cone-flat surfaces;
- Static loads or vibrations which tend to separate the ball from the vee–cone–flat surfaces. Springs are often used to provide a force holding the balls against the vee–cone–flat surfaces, but the spring force must be sufficient to prevent separation during static or vibration loading.

Entire instruments are sometimes mounted kinematically—not just optical components. Because the weight of an instrument can be large, this introduces excessive contact stresses, reducing the utility of the mount. In such cases, a *semi-kinematic mount* is a better option.

A semi-kinematic mount is one which relies not on point or line contact, but on the support provided by a small contact area. For example, a tripod is an ideal kinematic mount if the legs have spherical surfaces in contact with a ball-cone-flat. For some instruments—which must be bolted to their base, for example—this is not possible, and small feet must be used instead (Fig. 9.4a). Because of the area of the feet, this is now a semi-kinematic mount.

With three legs, a moment arm can be applied over the dimension b of the foot—for example, if the heel of the foot is on the floor, and the toe is raised off the floor due to machining errors (i.e., flatness or angularity). This moment can bend the support leg, which may in turn be transmitted to the instrument being supported by the tripod. In comparison, a non-kinematic four-leg approach is sometimes used, and this has an inherent "rocking" misalignment that can occur due to the fact that four points do not define a plane for the feet to sit on. In addition, very large moments can be applied through the base of the instrument (dimension B in Fig. 9.4b), which is more likely than the three-leg design to create structural deflection and optical misalignment within the instrument, or of the instrument with respect to others [3].

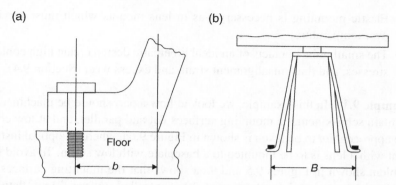

FIGURE 9.4 A semi-kinematic three-leg design with a small foot dimension as in (a) reduces the strain on the instrument in comparison with the non-kinematic four-leg design shown in (b). Adapted from Donald H. Jacobs *Fundamentals of Optical Engineering*, McGraw-Hill (1943).

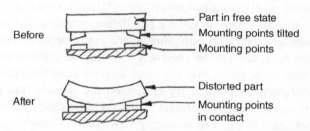

FIGURE 9.5 Mounting surfaces that are not in the same plane (i.e., not "coplanar") result in optical surface distortions. Credit: Dan Vukobratovich, SPIE Short Course SC014.

In general, when is an ideal kinematic mount *not* required? As three geometric points define a kinematic plane, the closer the size of the area is to an ideal point, the closer to kinematic the design is, and the larger the contact stresses will be. So the balance between system requirements and kinematic performance is best addressed with the use of design rules based on the mounting strain and distortion of optical surfaces, with a kinematic mount not required when

- Mounting surfaces are very flat and in the same plane ("coplanar")—see Figure 9.5;
- Optics or mounting structures are very stiff, and bolting through them will not strain the optic "much" if the mounting-surface coplanarity is poor;
- Mechanical dimensions are small, so deflections and strains from a flexible package will not magnify over a large distance;

- Elastic mounting is necessary—as in lens mounts which must survive over temperature;
- The small point contacts of an ideal kinematic design create high contact stresses, and thus misalignment strain and excess wear (Section 9.4).

Example 9.1   In this example, we look at how a part should be machined to maintain semi-kinematic mounting surfaces flat and parallel, and at low cost. The approach *not* to be taken is shown in Figure 9.6a, where an optical instrument with a lens is to be mounted to a baseplate with low strain. To avoid the problem shown in Figures 9.5 and 9.6a shows that the mounting surfaces of the instrument and baseplate must be controlled to very low flatness ($0.0005'' = 12.7\,\mu m$) over a large area (e.g., $100\,mm \times 100\,mm$). These are both expensive parts to machine, and by no means necessary for a low-strain semi-kinematic mount.

An alternative is shown in Figure 9.6b, where a reasonable surface flatness of $0.0005''$ over small mounting tabs ($10\,mm \times 10\,mm$) is much less expensive to machine. A requirement on surface parallelism is also added to insure the mounting surfaces—ideally, only three—are coplanar to within $0.0005''$. Similarly, the baseplate can be specified with the same tolerances over the small mounting areas, rather than the much larger area of the entire plate.

If the instrument is particularly stiff, then the flatness and parallelism requirements can be loosened to $0.001''$ (let's say), and this is a common trade in semi-kinematic design. The mounting tabs must, of course, also be stiff enough to prevent excessive motion of the instrument during vibration.

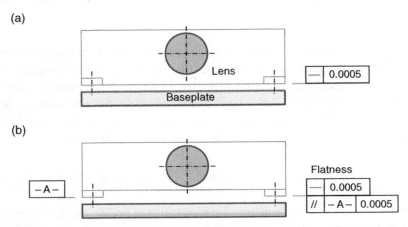

FIGURE 9.6 Semi-kinematic mounting of an instrument to a baseplate is much less expensive when maintaining flatness and parallelism requirements over a small mounting area, as in (b).

A similar trade is seen in another approach to semi-kinematic mounting, namely, the use of flexures, that is, thin beams that allow a small degree of bending in one direction but are very stiff in the others. An example is shown in Figure 9.7, where some degree of flexibility allows external deflections to be isolated from the optics by straining the flexure, rather than the instrument or the optics inside; tight requirements on surface flatness and parallelism are thus not necessary. Unfortunately, while stiffer flexures have a higher natural frequency and are more robust against damage from structural loads, they are also less kinematic in character, that is, they strain the instrument more than flexible beams. If such strain becomes excessive, it will be transmitted to the optics to the point where the WFE requirements of the system may be exceeded.

On occasion, attempts are made to use semi-kinematic mounting which, in practice, over-constrains the optic in a non-kinematic manner. An example is shown in Figure 9.8, where a receiver assembly is semi-kinematically mounted

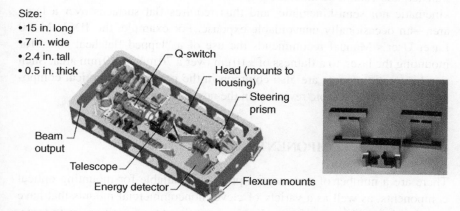

Size:
• 15 in. long
• 7 in. wide
• 2.4 in. tall
• 0.5 in. thick

Q-switch

Head (mounts to housing)

Steering prism

Beam output

Telescope

Energy detector

Flexure mounts

FIGURE 9.7 Semi-kinematic flexure mounts allow external deflections to be isolated from the optics. Credit: Floyd Hovis, Proc SPIE, Vol. 6100 (2006).

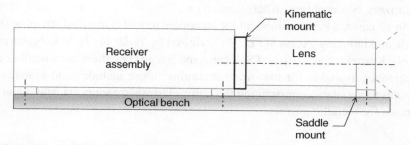

Kinematic mount

Receiver assembly

Lens

Optical bench

Saddle mount

FIGURE 9.8 A side view of the instrument shown in Figure 9.6. A saddle mount has been added to the lens to illustrate the concept of non-kinematic over-constraint.

to an optical bench. In addition, the lens is kinematically mounted to the receiver assembly (details not shown), and *also* supported with a saddle on its end to "prevent" sag of the lens assembly. This is a mistake; the lens is over-constrained by the saddle, and will be torqued by the following:

- The machining and assembly tolerances of the saddle, requiring shims to avoid;
- The long lever arm for optical bench deflection to bend the lens tube;
- Differences in thermal expansion and contraction between the support points, as well as thermal distortion of the bench applying a bending load.

The correct solution in this case is to stiffen the kinematic mount, and not attempt to "bypass" it with an additional constraint.

Finally, some instruments require lots of contact surface area for heat transfer, and cannot be mounted on small pads or feet. This is clearly neither kinematic nor semi-kinematic, and thus requires flat surfaces over a large area—an occasionally unavoidable expense. For example, the JDSU M110 Laser User's Manual recommends the use of a "lapped flat heat sink" for mounting the laser, to a flatness of $\pm 10\,\mu m$ over a 70 mm × 110 mm area. Such non-kinematic mounts are also common in the mounting of optical components themselves, a topic reviewed in the next section.

9.2 OPTICAL COMPONENT MOUNTS

There are a number of commercial products available for mounting optical components, as well as a variety of clever noncommercial mounts that have been used over the years for applications ranging from small optics to large. The emphasis in this book is on small-to-medium sized optics <100 mm or so in diameter. "Optics" includes lenses, prisms, mirrors, windows, laser crystals, polarizers, beamsplitters, filters, and so on.

In all cases, a key requirement for an optical mount is to avoid straining the optic and inducing excess WFE.[1] As reviewed by Yoder [4–7], Vukobratovich [8, 9], Kasunic et al. [10], and Schwertz and Burge [11], there are a number of techniques available for low-strain mounting; these include semi-kinematic mounts to avoid overconstraining the optic, elastomeric supports, and integral threads for metallic mirrors.

---

[1] Alignment requirements such as decenter may also need to be met, even though the mount itself may also be adjustable. If such adjustments can be avoided, then the design will typically be more robust against structural and thermal changes.

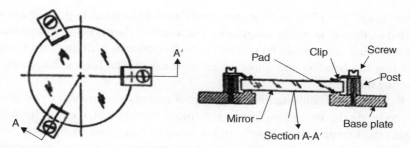

FIGURE 9.9 Mounting a mirror using three semi-kinematic pads and retaining clips. Credit: Paul R. Yoder Jr., *Mounting Optics in Optical Instruments*, SPIE Press (2008).

(a)   (b)

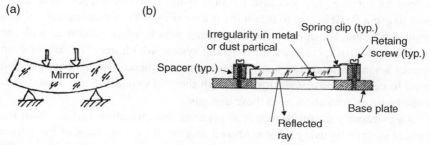

FIGURE 9.10 In (a), clamping forces that are offset from the mounting surfaces induce bending distortion and WFE. In (b), the WFE can be avoided by using three small mounting pads as shown in Figure 9.9. Credits: (a) Dan Vukobratovich, SPIE SC014 Course Notes; (b) Paul R. Yoder Jr., *Mounting Optics in Optical Instruments*, SPIE Press (2008).

Mounting an optic generally requires something simpler than vee–cone–flat arrangements. As shown in Figure 9.9, optics are often mounted on three small pads (e.g., Kapton tape). This three-legged chair concept—also known as a three-point mount—establishes a mounting plane. The pads are not actually points, but have a finite size, so the mount is semi-kinematic. In addition, as there are only three constraints, there are still three unconstrained DOFs: translation in the plane of the optic—that is, decenter which is potentially a serious problem for powered optics—and rotation of the optic about its centerline (rotor), an issue for cylindrical optics.

An important detail in the mounting of optical components—a necessity for avoiding component strain and associated WFE—is that the pad on which the optic sits be axially aligned with the clip or retainer clamping the component (Figs. 6.7 and 9.10). Mirror mounts are particularly critical because any surface deformation produces a 2× (reflective) effect on wavefront error. The three-"point" semi-kinematic mount must therefore be used; to provide a controlled method of avoiding irregularities or a lack of flatness over a large area, the mounting area should be relatively small compared with the mirror, thus avoiding the bending strain associated with misaligned forces that create a bending moment.

To avoid these deformations, some mirror mounts also use soft nylon locators to retain the mirror, avoiding glass–metal contact which can deform (or break) the mirror. Unfortunately, nylon tends to creep and change size with humidity, so this is not a good solution for precision mounts of optics with power or used outside a climate-controlled laboratory.

Mounting pads may be as simple as three small pieces of Kapton tape— useful for isolating the optic from the surface-finish irregularities of the metal mounting surface [12]. Because the tape is too thin to compress much, it is best to use a flexible clip to retain the mirror (Fig. 9.9); a disadvantage of this approach is that the clip is a flexible spring which, when combined with the mass of the mirror, creates a mass-spring system which may "rattle" (or even break) when vibrated at or near the resonance frequency. Radial pads are also used to center a lens or mirror, although thermal expansion and contraction must be taken into account in these designs.

An extremely useful design is to separate the mounting surface from the optical surface by using what is known as a "stalk" on the back of the mirror (Fig. 9.11). If the stalk is attached with low strain, this isolates the radial mounting forces from the optical surface, and minimizes surface deformation. Such stalks can be either integral (when the optic can be machined) or attached (when it cannot).

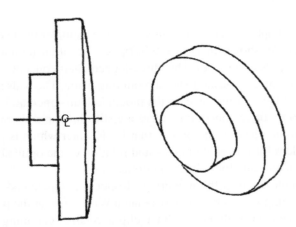

FIGURE 9.11 A small stalk on the back of a mirror isolates the radial clamping strain from the mirror surface. Credit: Dan Vukobratovich, SPIE SC014 Course Notes.

Another design useful for metallic mirrors is to isolate the mounting surface from the optical surface by machining mounting threads on the back of the mirror, separated from the mirror surface with slots (Fig. 9.12). The key concept is to again isolate the mounting forces from the optical surface by not allowing a direct path from the forces to the surface. Care is required with fabrication, however, as machining the slots after mirror polishing will release internal stresses, deforming the surface and creating WFE, while machining the slots before polishing allows the surface to deform under polishing loads; a match-machined pad in the slots during polishing is typically sufficient.

Many lenses are mounted using simple threaded retainers, holding the lens against a flat surface known as the *seat* (Fig. 9.13a). As we saw in Section 6.1.3, the retainer should have some degree of flexibility to avoid straining the lens. In addition, there is nothing "kinematic" about this design, so the seat and the lens surface must also be kept flat to minimize mounting strain and WFE. Care must also be taken that the decenter requirements are not exceeded, as can occur when there is excessive clearance between the lens and the housing diameters (see Chapter 4).

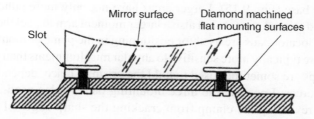

FIGURE 9.12 Mounting threads on the back of a mirror can isolate the mirror surface from the mounting forces. Credit: J. Zimmerman, Optical Engineering, Vol. 20, No. 2 (1981).

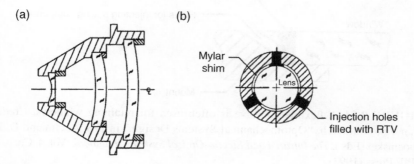

FIGURE 9.13 Both (a) retaining rings and (b) adhesives are used to mount optical components. Credits: (a) Dan Vukobratovich, SPIE SC014 Course Notes; (b) Paul R. Yoder Jr, *Mounting Optics in Optical Instruments*, SPIE Press (2008).

Becoming more and more common is the use of an adhesive—such as an epoxy or a flexible RTV (room-temperature vulcanizing) adhesive—to bond the lens in its housing (Fig. 9.13b). As was seen in Chapter 4, complex multielement lens assemblies often contain lenses bonded in individual lens cells. This may require a method to monitor and adjust lens-to-cell alignment (centering) before bonding. The cells are then press fit into the housing, with one (or two) critical lenses adjustable for final alignment of desired image quality.

In addition to mirrors and lenses, both windows and prisms require special techniques for mounting. Image quality is relatively insensitive to window deformation, and the requirement on mounting is usually that of avoiding breakage, not excess strain. As with lenses, a window mount often consists of a flexible adhesive bond around the entire circumference, allowing the window to expand and contract as its temperature changes (Fig. 9.14).

The primary optical requirement on mounting a prism—a relatively stiff optical component which cannot be easily strained—is similar to that of a window mount, that is, of avoiding breakage.[2] Prism mounts are usually semi-kinematic, in that the prism is bonded to its mount in three small, coplanar areas at its base (Fig. 9.15). Larger areas have not only more adhesion force and reduced contact stresses but also a larger moment arm to peel the adhesive. Kinematic locator pads are used for positioning in the plane of mounting base, as prisms are typically more sensitive to angular misalignments than positional. Simple clips are sometimes used to hold the prism in place, depending on the dynamic forces. Epoxy at the three mounting pads is also used in place of a clamp, preventing the clamp from cracking the sharp edges of the prism.

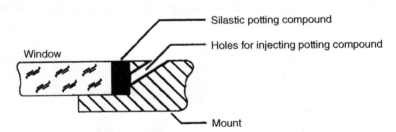

FIGURE 9.14 A flexible adhesive is often used for retaining windows. Credit: Dan Vukobratovich, "Optomechanical Systems Design," in J. S. Acetta and D. L. Shoemaker (Eds.), *The Infrared and Electro-Optical Systems Handbook*, Vol. 4, Chap. 3, SPIE Press (1993).

---

[2] The refractive index of a prism—and therefore its angle of refraction—can also depend on stress. See Ref. [13] for more details.

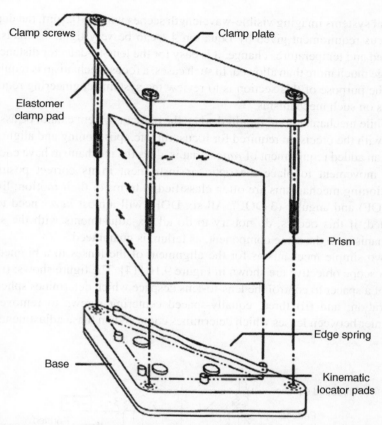

FIGURE 9.15 Both locator pads and mounting pads are used for low-strain mounting of a prism. Credit: Dan Vukobratovich, "Optomechanical Systems Design," in J. S. Acetta and D. L. Shoemaker (Eds.), *The Infrared and Electro-Optical Systems Handbook*, Vol. 4, Chap. 3, SPIE Press (1993).

As we have seen in Chapter 6, excess stresses from a clamp can also be avoided by chamfering or beveling the edges of the prism. Additional methods for mounting mirrors, lenses, windows, and prisms are covered in detail in Yoder [4–7], Vukobratovich [8, 9], and Kasunic et al. [10].

9.3 POSITIONING AND ALIGNMENT MECHANISMS

On occasion—despite the optomechanical designer's best efforts at avoiding them by using the methods reviewed in the previous sections—positioning and alignment mechanisms are required. For example, in the focusing of optical systems imaging visible-wavelength scenes with $\lambda \approx 0.5 \, \mu m$, the depth-of-focus requirement given by Equation 4.1 can be very tight. As loads are applied and temperatures change, it is easy for the lens-to-detector distance to change much more than allowed. In such cases, a focus mechanism is required, and the purpose of this section is to review the systems engineering requirements on such mechanisms.

While mechanisms are required when the optic or system cannot be assembled with the precision required for focus, tilt, etc., positioning and alignment have an added requirement of *range*, that is, does the mechanism have enough total movement to place the optic or instrument in its correct position? Positioning mechanisms are often classified in terms of their motion: linear (3 DOF) and angular (3 DOF). All six DOFs will almost *never* need to be aligned; if this occurs, do not try to do all six adjustments with the same mechanism on the same component, as failure is guaranteed.

Two simple mechanisms for the alignment of the lenses in a biomedical microscope objective are shown in Figure 9.16 [14]. The figure shows: (i) the use of a spacer to control the lens-to-lens despace which determines spherical aberration; and (ii) three, equally-spaced centering screws to remove the decenter between lenses which determines coma. When these adjustments are

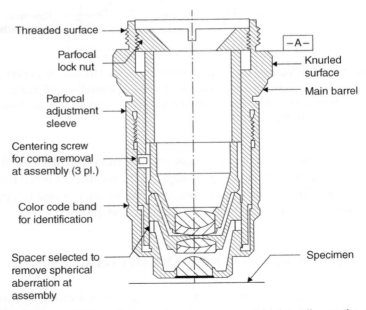

FIGURE 9.16 Lens-to-lens alignment for a microscope objective relies on the use of a spacer for despace and three centering screws for decenter. Adapted from J. R. Benford [14].

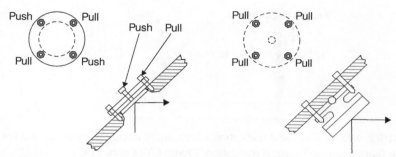

FIGURE 9.17 "Push-pull" and "pull-pull" alignment mechanisms are common but tend to drift. Adapted from John S. Chudy "The Optic as a Free Body," Photonics Spectra, Aug. 1985.

complete, a lock nut is used to tighten the assembly. Lens tilt is minimized through the perpendicularity of the mounting surfaces.

Another mechanism for the angular alignment of mirrors is the "push-pull" mechanism shown in Figure 9.17 [15]. This mechanism is a bit unstable, depending as it does on: (i) screws that must be adjusted by exactly the same amount to maintain the forces holding the mirror in place; (ii) the resulting stresses on the mount, and (iii) the drift (creep) that occurs from the load on the screw threads. When such a mechanism—or the related "pull-pull" mechanism—is required, the screws must be stabilized ("tied down") after adjustment with an epoxy or a removable adhesive such as those available from Glyptal, Inc. A better option for a manufacturing environment is to replace the screws with shims or epoxy, thus insuring a stable mount whose alignment cannot be easily changed—although at the cost of a potentially time-consuming alignment procedure.

Another common angular positioning mechanism—seen previously in Figure 9.2—is the kinematic mount. One drawback of this mechanism is the coupling of motions: an angular adjustment of the mirror also results in a displacement of the mirror surface. Another disadvantage of this mechanism is the springs that pull the alignment screws toward the kinematic locators (cone, vee, and flat). If these springs are not stiff enough, the alignment may change during vibration; too stiff, and the lifetime is limited by the wear or creep resulting from high contact stresses.

Flexures are also a commonly used as alignment mechanisms. *Flexures* are thin beams, bent by the application of a force (e.g., from an alignment screw). Both angular and linear adjustments are possible, though there is often coupling between the motions (Fig. 9.18). Flexures have the advantage of stability, precision, and repeatability. Their disadvantages include: (i) a small range of motion; (ii) coupling between angular and linear adjustments, so adjustments are not a

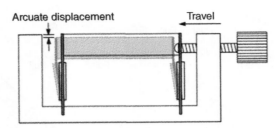

FIGURE 9.18 A horizontal translation with a linear flexure mechanism also results in tilt (not shown) and vertical translation. Credit: CVI Laser, LLC.

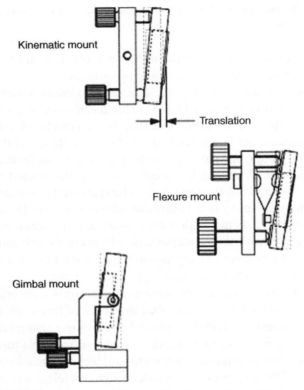

FIGURE 9.19 The gimbal mount rotates a mirror about its reflecting surface, avoiding the displacement of the kinematic and flexure mount. Credit: CVI Laser, LLC.

pure rotational motion; and (iii) heat transfer across thin flexures is difficult, and may lead to large temperature gradients that cause changes in alignment.

To avoid the coupling between axes found in kinematic and flexure mounts, a gimbal arrangement is used. *Gimbal* mechanisms allow pure rotation about a surface such as the front surface of a mirror (Fig. 9.19). This is obtained by

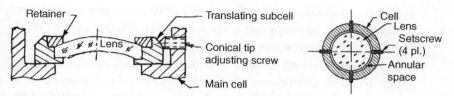

FIGURE 9.20 Adjustment screws are often used to control decenter. Credits: (a) Dan Vukobratovich, SPIE SC014 Course Notes; (b) Paul R. Yoder Jr., Mounting Optics in Optical Instruments, SPIE Press (2008).

using high-precision bearings whose center of rotation is the surface of the mirror. There are of course small displacements even in high-precision bearing rotation, but these will be much smaller than kinematic or flexure mounts.

Example 9.2 A simple alignment mechanism consists of adjusting screws that are used to center an optic before bonding (Fig. 9.20). What is the smallest increment (resolution) that the screw can position the optic?

A #4-40 screw [40 threads/in. (TPI)] is selected as a starting point. There are therefore $1/40''$ per thread, or $0.025''$ ($= 635 \,\mu m$). One revolution of the screw will therefore move the subcell $635 \,\mu m$.

The smallest possible motion is of course obtained by turning the screw less than a full revolution. A common number for human control of the screw is $1°$. Anything smaller is difficult for people to adjust with any repeatability. The minimum possible motion ("translation") of the subcell is therefore $\Delta z = 635 \,\mu m \times (1/360) = 1.76 \,\mu m$.

It is unlikely that this is sufficiently small for a precision lens assembly. Smaller screws with finer resolution are available, for example, a #2-56 with 56 TPI, or $1/56''$ per thread $= 0.0179''$/thread $= 454 \,\mu m$/thread. Additional solutions include micrometers with differential threads, or a geared positioner with better angular resolution than the human limit of $1°$.

Alignment mechanisms typically have a number of performance requirements that must be met. These requirements will flow down to specifications on the total travel range, resolution, load capacity, positioning errors due to mechanical tolerances, and so on. A list of such specifications will usually include the following:

- Degrees of freedom—Up to six DOFs may be required for each mechanism: three linear and three rotational. It is tempting to be a bit paranoid and use all six DOFs for each optic and every instrument. *This paranoia is almost guaranteed to fail—the system will be impossible to align, and unstable during vibration and temperature changes.* Use only as many DOFs as required—and that is often fewer than expected.

- Travel range—The total distance or angle the mechanism can move. This will be based on a worst-case optical and mechanical tolerance analysis. A large range does not typically have the best possible resolution (range vs. resolution trade).

- Resolution—The smallest distance or angle the mechanism can move.

- Stability—The change (due to drift, creep, vibration, or thermal changes) in a mechanism's position or angle over time.

- Linear positioning errors may result in cross-coupling of (i) the linear motion of one axis into the other two axes (flatness and straightness errors); and (ii) coupling into rotational motion such as pitch, roll, or yaw (Fig. 9.21). When there is no cross-coupling, the axes are said to be perfectly independent or "orthogonal."

- Rotary motion positioning errors may result in (i) wobbling due to tilt of the rotating part about the rotation axis; and (ii) eccentricity due to offset of the rotating surface from the rotation axis. Due to machining errors, even a rotary mechanism as simple as a screw thread has both of these errors to some degree (predominantly wobble).

In addition to these specifications, there are also the more general requirements of accuracy and precision. *Accuracy* refers to how close the mean value of a number of adjustments is to the ideal. *Precision*—also known as *repeatability*—is the standard deviation about the mean.

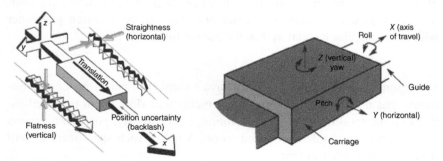

FIGURE 9.21 Linear positioning errors are a result of inevitable tolerances in the machining and assembly of mechanical components. Credit: (a) CVI Laser, LLC; (b) Permission to use granted by Newport Corporation; all rights reserved.

A lack of precision—a large variance in angular position, for example—may be due to either backlash or hysteresis in the mechanism. Backlash is the lack of motion when a mechanism is reversed. A classic example is a gear system, where mechanical tolerances between gears lead to a lack of motion when the gears change direction. It is also known as dead zone, looseness, slop, and free play. Preload can reduce it significantly. A mechanism with hysteresis does not return to its starting position after one or more cycles. It may be due to friction, stiction, temperature changes and gradients that cause parts to change position, material properties (e.g., PZT's), wear, and so on. The next section looks in more detail at the key material properties for mounts and mechanisms such as hardness and wear.

9.4   MATERIAL PROPERTIES AND SELECTION

While material properties such as stiffness and thermal expansion are as important for mechanism and mounts as they are for other optomechanical components, a unique aspect of kinematic and semi-kinematic design is the small contact area and potential for high contact stresses. In addition to contact-stress deformation (an elastic modulus effect), an additional damage that occurs is material being sheared off both surfaces when two moving surfaces are in contact, that is, the materials *wear* as a mechanism moves and changes position or angle. Hard materials—diamond, sapphire, and hardened steels—are less susceptible to surface damage and scratches from wear. Soft materials such as aluminum, copper, and brass, however, are much more susceptible. Wear "tracks" prevent a mechanism from returning to its original position, reducing the precision and repeatability of the device.

Measurement scales have been developed for quantifying material hardness. New scales were added as additional materials were measured with a hardness that exceeded the current scales. Scale resolution is also important, so different scales are used for different classes of materials with largely different hardness (e.g., glasses, metals, and plastics)—see Figure 9.22.

An additional material selection issue that is seen in mount and mechanism design is that of *galling*. This occurs when moving surfaces in contact are made of the same material; the surfaces then tend to adhere to each other, causing the mechanism to stop moving ("lock up"). For example, the threaded retainer shown in Figure 9.13a will require a huge force to turn if the retainer and mount are both made of aluminum. Galling does not occur for all materials, but

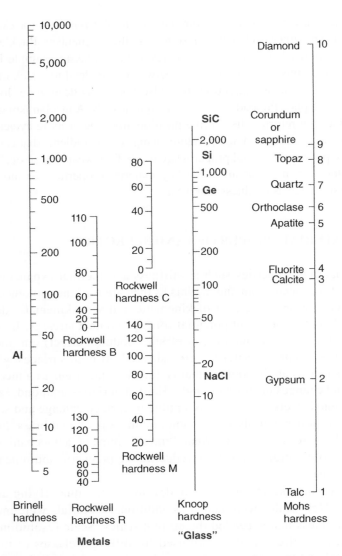

FIGURE 9.22 Different hardness scales are used to quantify material wear. Adapted from Paul R. Yoder Jr., *Opto-Mechanical Systems Design* (3rd Edition), CRC Press (2005).

aluminum-on-aluminum, stainless-on-stainless, and titanium-on-titanium are particularly susceptible. Different materials may also be prone to galling, for example, different types of stainless steel (17-4 PH on 440C) also gall. To avoid galling, many commercial mounts use threaded brass inserts with stainless steel alignment screws [1].

PROBLEMS

9.1 Does it matter what angle is used to machine the "vee" in vee–cone–flat or three-vee kinematic mounts? Is 60°, for example, preferred over 45° or 30°?

9.2 Estimate the flatness of the mounting surfaces needed to maintain the tilt angle of the optical instrument shown in Example 9.1 to <10 microradians. The base of the instrument is 100 mm × 100 mm in size; also estimate the flatness if the system is 1 m in size.

9.3 In Example 9.2, we saw that a #4-40 screw thread may be inadequate for given decenter requirements. For a decenter requirement of ±1 micron, is a #2-56 screw thread acceptable?

9.4 A focus adjustment mechanism is designed to produce a linear displacement of 10 μm for a known voltage change. The focus change for ten displacements was measured to be: 10.6, 10.3, 10.3, 9.9, 10.3, 10.2, 10.5, 10.6, 10.3, and 10.3 μm. What is the accuracy of the measurement, expressed as a percentage? What is the precision (repeatability)? Is there a bias in the design that might be avoided with a redesign of the mechanism?

REFERENCES

1. S. Cabatic and K. D. Li Dessau, "Designing cost-effective precision optical mounts," *Optics and Photonics News*, May 2000, pp. 30–34.
2. P. Giesen and E. Folgering, "Design guidelines for thermal stability in opto-mechanical instruments," Proc. SPIE, Vol. 5176, pp. 126–134 (2003).
3. D. H. Jacobs, *Fundamentals of Optical Engineering*, New York: McGraw-Hill (www.mcgraw-hill.com) (1943).
4. P. R. Yoder, Jr., *Mounting Optics in Optical Instruments* (2nd Edition), Bellingham: SPIE Press (www.spie.org) (2008).
5. P. R. Yoder, Jr., *Opto-Mechanical Systems Design* (3rd Edition), Boca Raton: CRC Press (www.crcpress.com) (2005).
6. P. R. Yoder, Jr., "Mounting optical components," in M. Bass, E. W. Van Stryland, D. R. Williams, and W. L. Wolfe (Eds.), *Handbook of Optics* (2nd Edition), Vol. 1, New York: McGraw-Hill (www.mcgraw-hill.com) (1995), Chap. 37.
7. R. E. Fischer, B. Tadic-Galeb, and P. R. Yoder, Jr., *Optical System Design* (2nd Edition), New York: McGraw-Hill (www.mcgraw-hill.com) (2008), Chaps. 16–18.
8. D. Vukobratovich, *Introduction to Optomechanical Design*, SPIE SC014 Course Notes (www.spie.org) (2009).

9. D. Vukobratovich, "Optomechanical systems design," in J. S. Acetta and D. L. Shoemaker (Eds.), *The Infrared and Electro-Optical Systems Handbook*, Vol. 4, Bellingham: SPIE Press (1993), Chap. 3.

10. K. J. Kasunic, J. Burge, and P. R. Yoder, Jr., *Mounting of Optical Components*, SPIE SC1019 Course Notes (www.spie.org) (2013).

11. K. Schwertz and J. H. Burge, *Field Guide to Optomechanical Design and Analysis*, Bellingham: SPIE Press (2012).

12. K. J. Kasunic and P. B. Forney, "Design of a rotary indexing mechanism for a cryogenic-vacuum environment," Proc. SPIE, Vol. 973 (1988). doi:10.1117/12.948389.

13. K. J. Kasunic, *Optical Systems Engineering*, New York: McGraw-Hill (www.mcgraw-hill.com) (2011).

14. J. R. Benford, "Microscope objectives," in R. Kingslake (Ed.), *Applied Optics and Optical Engineering III*, New York: Academic Press, pp. 145–182 (1965), Chap. 4.

15. J. S. Chudy, "The optic as a free body," *Photonics Spectra*, August 1985, pp. 49–59.

10

SYSTEM DESIGN

After many late nights of trade studies, design, analysis, assembly, and testing … of working with vendors to fabricate lenses, machine parts, and order components … of overextended schedules and overspent budgets—a well-earned miracle will finally come together in the form of working hardware (Fig. 10.1), reminding us why we are not Viewgraph engineers. "We," of course, includes not just the optomechanical engineer, but the optical, electrical, software, and systems engineers—all key players in the success of the project.

In addition to an experienced team, some hardware projects will also require analysis tools that are more accurate than the rules-of-thumb and back-of-the-envelope equations used to develop a first-order design. Unfortunately, it is thought by some that modern "back-of-the-elephant" software tools eliminate the need to think! Just plug in some numbers, and away ya go, eh. Fortunately *NOT*—with the complexity of the structures that can be analyzed with software, *it's more necessary than ever to ask: "Is this result reasonable?"* The most important tool of optomechanical systems engineering is therefore: The *SIMPLE* Model.

The simple model usually leads to simpler designs: more reliable, lower cost, longer lifetime, and so on. A good system designer—optomechanical or

Optomechanical Systems Engineering, First Edition. Keith J. Kasunic.
© 2015 John Wiley & Sons, Inc. Published 2015 by John Wiley & Sons, Inc.

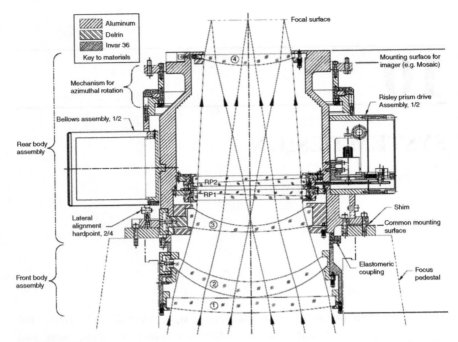

FIGURE 10.1 An assembly drawing of an optomechanical system shows the relationships between individual components. Credit: George H. Jacoby et al., Proc SPIE, Vol. 3355 (1998).

otherwise—also uses simple models to look for connections and interrelationships that are not immediately obvious:

- While a strain-free bond thickness can be found for a lens whose temperature is not changing, what is the natural frequency of the lens-plus-bond system?
- What are the effects of CTE inhomogeneities on thermal stress?
- What are the effects of putting a cooling fan directly on an optical surface such as the back of a primary mirror?
- What is the natural frequency of a spring-loaded kinematic mount?
- What are the simultaneous effects of vibration and temperature changes on window defocus?

Despite the obvious benefits of the simple model, not all optical systems are simple enough to be accurately analyzed using the order-of-magnitude equations in this book. The calculation of thermal distortion, natural frequencies, and the like can be extremely complex, and back-of-the-envelope estimates may be insufficient for anything other than first-order sizing. The results are useful to the optical systems engineer for purposes of initial design and trend analysis, but they are not always accurate enough to be used for developing component specifications such as tube diameter and material stiffness. Simply choosing a large diameter and stiff material may also be unacceptable in light of the additional constraints of system size, weight, and cost. This means that techniques are needed for the more accurate prediction of optical system performance, particularly when high-precision tolerances or high-risk systems are involved [1].

Accurate predictions of performance must therefore address two issues: complex geometries and the availability of integrated design tools that enable the simultaneous trading off of optical, thermal, and structural parameters. Complex geometries can be analyzed using finite-element analysis (FEA) and analytical cases from design handbooks such as *Roark's Equations for Stress and Strain* [2] and the *Handbook of Heat Transfer* [3]. Fully integrated design software is under development and is usually described as "structural, thermal, and optical performance (STOP)" analysis [4–6].

10.1   STOP ANALYSIS

As illustrated in Figure 10.2, STOP analysis consists of four components: (1) developing a computer-aided design (CAD) model of the optical and mechanical systems; (2) using finite-element models to calculate the temperature and strains of the system under thermal and structural loading; (3) converting the calculated effects into surface figure errors and wavefront errors; and (4) assessing the consequences of these errors using optical performance metrics such as blur size and modulation transfer function (MTF).

From the perspective of the optomechanical engineer, detailed system design begins with a computer-aided design (CAD) model of the optical, mechanical, and electro-mechanical components. CAD models are extremely sophisticated, with complex internal code to generate shapes, overlaps, shading, cross sections, and so on. An example screen shot of a 3D model is shown in Figure 10.3, where the software tools used to develop a CAD model are typically ProEngineer ("ProE") or SolidWorks.

Once a CAD model and assembly drawings are established, it is then necessary to insure that the thermal and structural environments do not induce excessive WFE. This is also shown in Figure 10.2, where the thermal analysis is completed before the structural, as thermal expansion and the temperature distribution can in part determine the structural deflections.

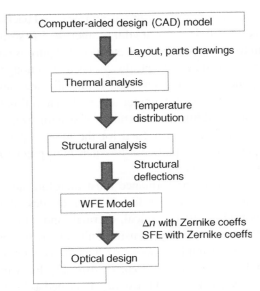

FIGURE 10.2 System design often requires the use of CAD, thermal analysis, structural analysis, WFE, and optical models.

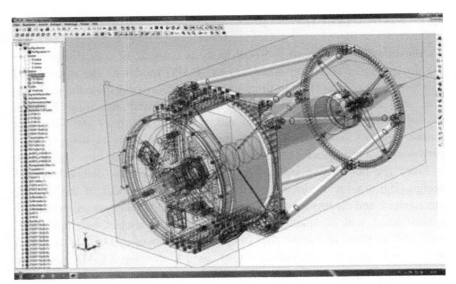

FIGURE 10.3 A typical CAD model and software GUI. Credit: Alluna Optics.

It is entirely possible to rely on the first-order analysis tools developed in this book for the thermal and structural analysis, but it is also possible the results will not have the fidelity or accuracy required to minimize the number of times prototypes need to be built. That is, there is in general a trade between the time

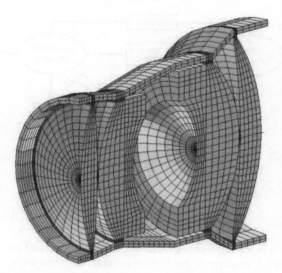

FIGURE 10.4 Finite-element analysis (FEA) of a lens requires that the lens be subdivided ("meshed") into small elements. Credit: Sigmadyne, Inc. http://sigma-dyne.com/.

spent analyzing a system and the amount of time building and testing it. On occasion, there are people on a project with enough experience and "intuition" to reduce the amount of time spent on analysis. But without such people—*and once the simple, first-order tools in this book have been utilized*—the purpose of higher order analysis is to reduce the cost and time spent on the "build and test" component; a higher accuracy model—if not too much time is spent developing it—allows this.

The higher accuracy models are developed using finite-element analysis (FEA). Finite-element models and finite-element analysis are used for geometries that are not easily broken down into the analytical cases found in design handbooks. This occurs less often in thermal than in structural design, where the tensor nature of stress and strain requires geometries such as cones, tubes, struts, and so on to have a large (though finite) number of spatial components. An example is shown in Figure 10.4, where the entire volume of a lens–cell structure has been divided ("meshed") into small elements for which temperature, stress, and strain are computed separately; in turn, these factors determine the tilt, decenter, and defocus as well as optical component distortions and thermo-optic and stress-optic effects. The selection of enough elements to give accurate results within a reasonable computation time is a critical skill of the thermal and structural analysts.

STOP analysis using finite-element analysis can take a number of paths. The simplest technique for integrated design is to connect the structural

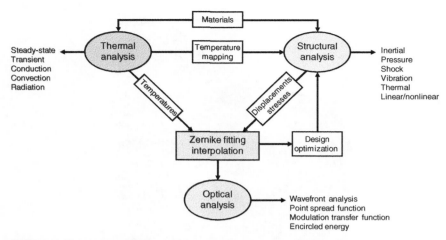

FIGURE 10.5 STOP analysis brings together elements of optical, thermal, and structural design to generate accurate predictions of total wavefront error. Credit: Keith B. Doyle, Victor L. Genberg, and Gregory J. Michels, *Integrated Optomechanical Analysis* (2nd Edition), SPIE Press (2012).

performance (Chapters 5, 6, and 7) and thermal performance (Chapter 8) with the optical performance; as suggested by Figure 10.5, the best accuracy is obtained by simultaneously including all the relevant thermal and structural effects and interactions. In the current state of the art, STOP analysis calculates temperature and stress distributions to determine the element displacements and wavefront error; additional structural effects (e.g., self-weight, vibrations due to motors and structural motion) can also be included.

Software packages for thermal FEA include Thermal Desktop, SINDA, and TSS. The output from the thermal model is a temperature distribution for individual lenses, assemblies, and mechanical structures. The temperature distributions are then fed into a structural analysis code such as NASTRAN, which along with the static and dynamic structural loads on the system determine the surface strains and element motions (tilt, decenter, and defocus).

Summarizing: the goal of an integrated STOP analysis is not to turn the structural designer into an optical engineer, or an optical designer into a thermal analyst. Rather, the intent is to reduce the barriers that prevent different designers from working together. Just as there are file-format and coordinate-system barriers between software packages, there are also difficulties in communication between designers with different areas of expertise. These difficulties may be greatly alleviated by optical systems engineers who have a good understanding of the relevant disciplines and can talk the same language as the structural, thermal, optical, and control system designers [7]. Such a collaborative approach can greatly reduce the risk and cost of system development [6]. The next section describes in more detail the optical component of STOP analysis.

10.2 WFE AND ZERNIKE POLYNOMIALS

The temperatures, stresses, deflections, and deformations determine the misalignments and surface figure errors of the optical components. NASTRAN and Abaqus do not calculate wavefront error from these parameters, so results from the structural models must be exported into a WFE model such as SigFit or equivalent proprietary software [8, 9]. As shown in Figure 10.6, SigFit uses the estimated structural, thermostructural (CTE), thermo-optic (dn/dT), and stress-optic ($dn/d\sigma$) effects to compute the change in curvature ("surface deformations") for each optical surface, misalignments ("rigid-body motions") such as tilt and decenter between components, and change in index of refractive materials. The temperature- and stress-dependent index

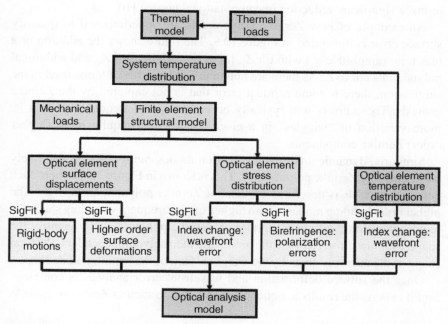

FIGURE 10.6 Physical effects which SigFit includes are rigid-body motions, surface deformations, and refractive index changes due to stress and temperature changes. Credit: Keith B. Doyle, Victor L. Genberg, and Gregory J. Michels, *Integrated Optomechanical Analysis* (2nd Edition), SPIE Press (2012).

$n(T, \sigma) = n_o + \Delta n_t + \Delta n_s$ for a nominal design index n_o, a change in index Δn_t due to thermo-optic effects, and a change in index Δn_s due to stress-optic effects.

Changes in surface curvature and thermo-optic and stress-optic wavefront errors are often fitted to polynomials originally developed by Fritz Zernike. These polynomials can be used to generate an interferogram mapping of the WFE [1]; this mapping is exported to optical design codes such as Zemax or Code V, which allow the effects on image quality to be evaluated. For imagers, the most important output from the optical code is WFE, from which image quality metrics (blur size, MTF, etc.) are determined [7].

It is possible to quantify surface deformations and wavefront errors with an arbitrary set of polynomials; for example, we could describe a wavefront shape as a sum $w(x,y) = \Sigma k_i p_i(x,y)$ for a weighting factor k_i and set of polynomials $p_i(x,y)$. The benefit of the Zernike polynomials $Z_i(r,\theta)$ shown in Figure 10.7 is that many components are of the same form as the aberrations used for the design and test of the system—for example, defocus, astigmatism, coma, and spherical aberration. In addition, deformations or WFE over thousands of grid points can be represented by the first fifty or so Zernike terms, a significant reduction in computational burden [10, 11].[1]

An example of how Zernike polynomials are superimposed to quantify surface error is illustrated in Figure 10.8. The figure shows the addition of a bias term (amplitude a_o) with tilt Z_1, focus Z_2, astigmatism Z_3, and additional polynomials out to Z_i. As there are not an infinite number of terms used in the summation, there is some residual error that is not captured by the Zernike analysis. These errors will typically be higher spatial frequencies—that is, more variation or "wiggles" in a given length—that require more (higher order) Zernike components.

Similarly, dynamic analysis of vibration modes can usually be accurately represented by Zernike polynomials. This is shown in Figure 10.9, where each bending mode Φ_i is described by a set of Zernike polynomials $Z_1 \rightarrow Z_n$. The higher order bending modes—with higher spatial-frequency bending shapes—require the use of sufficient number of Zernike polynomials to quantify the shape with low error. Because the modes are themselves independent, the total shape of the mirror is given by the weighted sum of the bending mode shapes.

Once the surface deformation and wavefront error analysis is complete, SigFit exports the results to optical design software such as Zemax or Code V.

---

[1] Zernike polynomials are also independent ("orthogonal") in r–θ coordinates over a circular aperture; they thus allow a decomposition of an arbitrary surface or wavefront shape into a sum ("linear superposition") of Zernike components. While the continuous polynomials $Z_i(r,\theta)$ are orthogonal, however, sampled (or digitized) representations of the polynomials will in general not be.

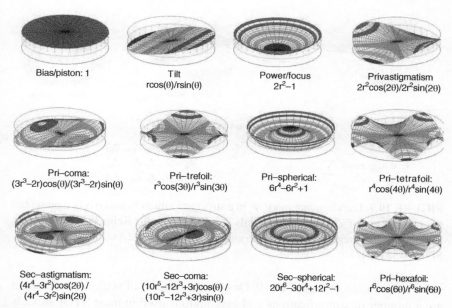

FIGURE 10.7 The first 12 of the Zernike polynomials used to describe surface shapes and wavefronts in a form that maps to optical aberrations. Credit: Keith B. Doyle, Victor L. Genberg, and Gregory J. Michels, *Integrated Optomechanical Analysis* (2nd Edition), SPIE Press (2012).

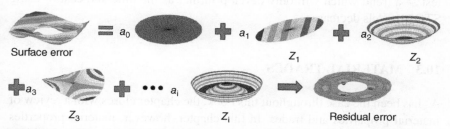

FIGURE 10.8 Illustration of the use of Zernike polynomials to represent a surface error. Credit: Keith B. Doyle, Victor L. Genberg, and Gregory J. Michels, *Integrated Optomechanical Analysis* (2nd Edition), SPIE Press (2012).

If required, the new optical prescription and CAD model are fed back into the FEA models, and the optomechanical design cycle starts again with thermal, structural, vibration, and wavefront error analysis.

Because the number of models used—and the potential for error—is large in STOP analysis, vendors have developed commercial software for optical sensor analysis that integrates the various STOP elements and allows more design options to be evaluated in a shorter time; a screen shot from such

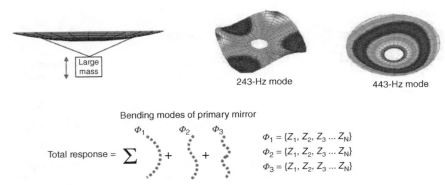

243-Hz mode

443-Hz mode

Bending modes of primary mirror

Total response = $\sum$

φ_1 φ_2 φ_3

$\varphi_1 = \{Z_1, Z_2, Z_3 \ldots Z_N\}$
$\varphi_2 = \{Z_1, Z_2, Z_3 \ldots Z_N\}$
$\varphi_3 = \{Z_1, Z_2, Z_3 \ldots Z_N\}$

FIGURE 10.9 Each bending mode $\varPhi_i$ in a structural vibration analysis is represented by a unique set of Zernike polynomials Z_1, Z_2, etc. Credit: Keith B. Doyle, Victor L. Genberg, and Gregory J. Michels, *Integrated Optomechanical Analysis* (2nd Edition), SPIE Press (2012).

software is shown in Figure 10.10. Despite the benefits of such software, there are a number of simplifications and assumptions used in most STOP models to prevent them from replacing—except for the most trivial of geometries or difficult of test environments—"build and test" as the ultimate standard for validation, verification, or product acceptance. Even so, the trend is toward using STOP analysis to reduce the often significant expense of "build and test"—a trend which will only develop further as the time and cost of using STOP models decreases.

10.3 MATERIAL TRADES

As has been the case throughout this book, the chapter closes with a review of material properties and trades. In this chapter, however, material properties from all sections—fabrication, alignment, structural, thermal, and kinematic—are brought together into the overall systems trades. For example it is sometimes necessary to stiffen a mounting structure to reduce its deflections; at the same time, stiffening can also lead to an increase in thermal distortion as the temperature changes, with the lighter optic (and the smaller thermal mass) changing size before the heavier mount, even if they are the same material. Such trades for structural materials—summarized in Table 10.1— can often lead to a re-direction of the design [12].

Another material trade is between the optical and mechanical properties of refractive materials. This trade is not usually pursued; it is often the lens designer who selects the material based on optical properties alone.

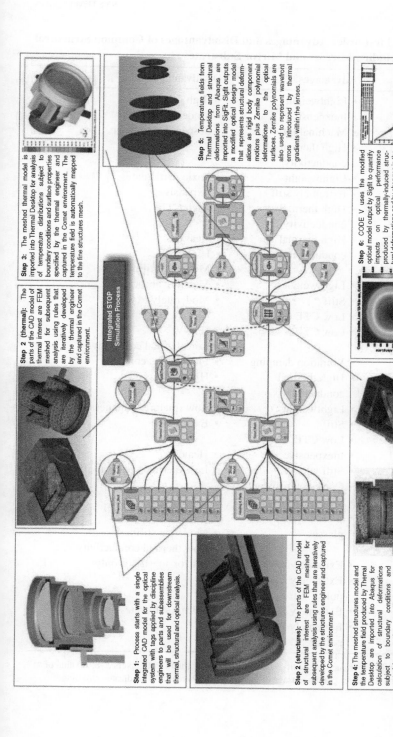

Step 1: Process starts with a single integrated CAD model for the optical system with tags applied by discipline engineers to parts and subassemblies that will be used for downstream thermal, structural and optical analysis.

Step 2 (structures): The parts of the CAD model of structural interest are FEM meshed for subsequent analysis using rules that are iteratively developed by the structures engineer and captured in the Comet environment.

Step 4: The meshed structures model and the temperature field produced by Thermal Desktop are imported into Abaqus for calculation of structural deformations subject to boundary conditions and material properties specified by the structures engineer and captured in Comet. Lens mount contact stresses are modeled accurately, as shown to the right.

Step 2 (thermal): The parts of the CAD model of thermal interest are FEM meshed for subsequent analysis using rules that are iteratively developed by the thermal engineer and captured in the Comet environment.

Step 3: The meshed thermal model is imported into Thermal Desktop for analysis of temperature distributions subject to boundary conditions and surface properties specified by the thermal engineer and captured in the Comet environment. The temperature field is automatically mapped to the fine structures mesh.

Step 5: Temperature fields from Thermal Desktop and structural deformations from Abaqus are imported into SigFit. Sigfit outputs a modified optical design model that represents structural deformations as rigid body component motions plus Zernike polynomial deformations to the optical surfaces. Zernike polynomials are also used to represent wavefront errors introduced by thermal gradients within the lenses.

Step 6: CODE V uses the modified optical model output by Sigfit to quantify impacts on optical performance produced by thermally-induced structural deformations and by changes in the refractive indices. An exit pupil wavefront error map is shown at left and a Modulation Transfer Function plot at right.

Integrated STOP Simulation Process

FIGURE 10.10 A screenshot from a commercial software tool for integrating the many components of STOP analysis. Credit: Comet Solutions, Inc.

TABLE 10.1 First-order Advantages and Disadvantages of Common Structural Materials, including Mirrors and Mirror Substrates

Material	Advantages	Disadvantages	Use
Aluminum (6061-T6)	• Lightweight • Inexpensive • Machinability • High thermal conductivity	• High CTE • Moderate stiffness • Susceptible to wear	1
Beryllium	• Lightweight • Extremely stiff • High thermal conductivity	• Toxic particulates • Expensive • High CTE	3
Copper (OFHC)	• High thermal conductivity	• Heavy • High CTE • Moderate stiffness	2
Graphite epoxy (GrE)	• Lightweight • Stiff • Low CTE (tunable)	• Moisture absorption • Moderately expensive	2
Invar	• Low CTE	• Heavy • Low thermal conductivity	2
Magnesium	• Vibration damping • High thermal conductivity	• Flammable, corrosive • High CTE	3
Silicon carbide (SiC)	• Lightweight • Stiff • Low CTE	• Low fracture toughness • Expensive	3
Stainless steel (304 CRES)	• Inexpensive • Stiff • Corrosion resistant • Wear resistant	• Heavy • Low thermal conductivity	2
Titanium (6Al-4V)	• CTE matches many glasses	• Moderate weight • Moderate stiffness • Low thermal conductivity	2
ULE	• Extremely low CTE • Extremely smooth polish	• Mirror substrates only	1
Zerodur	• Extremely low CTE • Extremely smooth polish	• Mirror substrates only	1

Higher level comparisons such as specific stiffness E/ρ can be found in Chapters 5–8. A value of "1" in the Use column indicates common usage of the material, whereas a "3" indicates rare usage.

TABLE 10.2 Comparison of Mechanical Properties of SCHOTT's SF57 and N-SF57 Materials

Property	SF57	N-SF57
Refractive index, n_d	1.8467	1.8467
Density, ρ	5.51 g/cm^3	3.53 g/cm^3
Elastic modulus, E	54 GPa	96 GPa
CTE, α_t (−30 C → +70 C)	8.3 × 10^{-6}/K	8.5 × 10^{-6}/K
dn/dT (absolute)	6 × 10^{-6}/K	−2.1 × 10^{-6}/K
Thermal conductivity, k	0.62 W/m-K	0.99 W/m-K

Unfortunately, this has led to situations where the mechanical properties that have been neglected end up controlling the design. For example, cemented doublets can break because of differential thermal expansion between the doublet materials and the cement. Another example is the selection of calcium fluoride as a window material because of its broadband optical transmission; as we have seen in Chapter 6, however, calcium fluoride also has low fracture toughness, and is easily susceptible to breaking when its surface has small cracks—as it likely will when it is used as a window.[2]

It is perfectly reasonable, then, for the optomechanical engineer to question the lens designer's choice of materials. This is especially the case when the newer lead-free materials are selected based on "equivalent" optical properties, or required due to the unavailability of the older materials. While it is true that the refractive index of such replacement materials is the same—compare SF57 with N-SF57, for example—the optomechanical properties are often quite different. This is shown in Table 10.2, where the density ρ, stiffness E, dn/dT, and even thermal conductivity k are drastically different! Such "street" knowledge can save many weeks of effort, with "lessons learned" summarized throughout this book for many other subtleties of optomechanical systems engineering as well.

Building systems that actually meet specifications requires more than a knowledge of optics and contemporary design codes. Excellence also comes from a kind of 'street' knowledge, learned not from textbooks but from experience, mostly from failures.

Anthony Smart, "Folk Wisdom in Optical Design," *Applied Optics,*
1 December 1994.

---

[2] If the window is static pressure loaded, the side of the window in tension is the side that is first susceptible to fracture; a vibration load will alternate between a tension and compression force on both sides of the window.

REFERENCES

1. K. B. Doyle, V. L. Genberg, and G. J. Michels, *Integrated Optomechanical Analysis* (2nd Edition) Bellingham: SPIE Press (2012).

2. W. C. Young and R. G. Budynas, *Roark's Formulas for Stress and Strain* (7th Edition), New York: McGraw-Hill (2001).

3. W. Rohsenow, J. Hartnett, and Y. Cho, *Handbook of Heat Transfer* (3rd Edition), New York: McGraw-Hill (1998).

4. J. Miller, M. Hatch, and K. Green, "Predicting performance of optical systems undergoing thermal/mechanical loadings using integrated thermal/structural/optical numerical methods," Opt. Eng., Vol. 20, No. 2 (1981).

5. J. D. Johnston, J. M. Howard, and E. Gary, "Integrated modeling activities for the James webb space telescope: structural-thermal-optical analysis," Proc. SPIE, Vol. 5487 (2004).

6. J. Geis, J. Lang, L. Peterson, F. Roybal, and D. Thomas, "Collaborative design and analysis of electro-optical sensors," Proc. SPIE, Vol. 7427 (2009).

7. K. J. Kasunic, *Optical Systems Engineering*, New York: McGraw-Hill (2011).

8. P. A. Coronato and R. C. Juergens, "Transferring FEA results to optics codes with Zernikes: a review of techniques," Proc. SPIE, Vol. 5176 (2003).

9. R. C. Juergens and P. A. Coronato, "Improved method for transfer of FEA results to optical codes," Proc. SPIE, Vol. 5174 (2003).

10. B. Qi, H. Chen, and N. Dong, "Wavefront fitting of interferograms with Zernike polynomials," Opt. Eng., Vol. 41, No. 7 (2002).

11. E. P. Goodwin and J. C. Wyant, *Field Guide to Interferometric Optical Testing*, Bellingham: SPIE Press (2006).

12. M. Schreibman and P. Young, "Design of infrared astronomical observatory (IRAS) primary mirror mounts," Opt. Eng., Vol. 20, No. 2 (1981).

INDEX

WILEY SERIES IN PURE AND APPLIED OPTICS

Founded by Stanley S. Ballard, University of Florida

EDITOR: Glenn Boreman, University of North Carolina at Charlotte

Printed and bound by CPI Group (UK) Ltd, Croydon, CR0 4YY

16/04/2025

14658343-0005